RON FOLMAN

THE HUMAN TEST

How Predictability,
Creativity, and the
Quantum Mind
Will Redefine Life
in the Age of AI

Prometheus Books
Essex, Connecticut

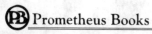

Prometheus Books

An imprint of The Globe Pequot Publishing Group, Inc.
64 South Main Street
Essex, CT 06426
www.globepequot.com

Distributed by NATIONAL BOOK NETWORK

Copyright © 2025 by Ron Folman

All rights reserved. No part of this book may be reproduced in any form or by any electronic or mechanical means, including information storage and retrieval systems, without written permission from the publisher, except by a reviewer who may quote passages in a review.

British Library Cataloguing in Publication Information Available

Library of Congress Cataloging-in-Publication Data Available

ISBN 9781493089208 (cloth) | ISBN 9781493089215 (epub)

∞™ The paper used in this publication meets the minimum requirements of American National Standard for Information Sciences—Permanence of Paper for Printed Library Materials, ANSI/NISO Z39.48-1992.

They say that I am not who I am—and it scares me,
For if I am not who I am—who am I?

—*Written in Yiddish by Abraham Goldfaden, 1840-1908*

Dedicated to my fellow humans.

To my parents, Ahuva (Gordon) Folman and Yeshayahu Folman, who planted the seeds of curiosity and love in my mind and nurtured them.

Contents

PREFACE ix
1. They Say That I Am Not Who I Am—and It Scares Me ix
2. Artificial Intelligence xv

INTRODUCTION 1
1. Eve 1
2. Going beyond Consciousness 6
3. Machine Death 8
4. A Connection to Consciousness 12
5. The Pill 15
6. The Book 17

PART I: THE PREDICTABILITY EXPERIMENT

Chapter 1: The Near Future Will Change Everything 27
1. You Can't Ignore the Results of a Well-Performed Experiment 27
2. Measuring Predictability 31
3. The Human Machine 44

v

PART II: IS DETERMINISM IN THE BRAIN REALLY POSSIBLE?

Chapter 2: Common Arguments against the Possibility of Predictability   61
1. Emergence versus Reductionism   61
2. Free Will   65
3. The Soul   76

PART III: RETHINKING THE OBVIOUS IS WORTH FIGHTING FOR

Chapter 3: The Hard Road to a New Understanding   85
1. Truth and Objectivity   88
2. Fighting Our Ego   92
3. Words Are Not Enough   98

Chapter 4: Enabling Our Mind to Make a Leap of Thought   103
1. The Thought Experiment   103
2. Our Brain in a Glass Vessel   109

PART IV: A NEW LIFE REQUIRES A NEW DEATH

Chapter 5: Going Down the Rabbit Hole   115
1. The Question   115
2. The Present Limits of Our Knowledge   119

Chapter 6: Predictability and a New Definition for Human Life and Death   123

PART V: THE FUTURE OF PREDICTABILITY

Chapter 7: Can We Undo Predictability?   135
1. Can Internal Noise Save Us?   135
2. The Quantum Mind   142
3. New Physics   154

Chapter 8: The Bottom Line of Predictability   163
1. A Simplified Model for How Our Brain Produces Outcomes   163
2. Conclusion   168

PART VI: CONSEQUENCES AND POSSIBLE PATHWAYS TO TRANSCENDENCE

Chapter 9: The Pill   179

1. War   186
2. Happiness   192
3. Creativity   199
4. Logic   202
5. Society   206

Epilogue   211

APPENDICES

Appendix A: The Brain   217

Appendix B: Attempts to Quantify Consciousness and Self-Awareness   221

Appendix C: Research Dealing with Prediction of Human Behavior   227

Appendix D: Kitsch and the Human Satisfaction Resonance   233

Notes   235

Index   253

Preface

The Human Test tells the story of the most profound experiment ever conducted on humans, starting to take place right now, and which, astonishingly, has been materializing mostly under the radar. It is about us, humans. It describes what fundamental new things we are about to learn about our species in the very near future, perhaps even within our lifetime. In this sense we are living in dramatic times. What changed? Our ability for introspection is about to be completely transformed by artificial intelligence (AI). This is an earthshaking aspect of AI rarely discussed. In parallel, our understanding of quantum mechanics, which governs atoms and molecules, the building blocks of our brain, is constantly improving. *The Human Test* connects these two topics. In this preface I briefly touch upon the topic of AI and the introspection experiment it is about to allow us. In the introduction I start speaking of the quantum realm. But before we begin our expedition, I would first like to tell you in a few words about my personal journey, which eventually led me to write this book.

1. THEY SAY THAT I AM NOT WHO I AM—AND IT SCARES ME
Physical survival is obviously high on everyone's agenda. In my family it was similar, but perhaps to an extreme. My father managed to survive the Auschwitz death camp, an inferno never before seen on Earth. My mother barely

escaped her burning city of Minsk as the Nazi army advanced, just to be in the crosshairs of Stuka dive-bombers, and was later caught by the British in a dramatic battle at sea when she tried to make it, together with other illegal refugees, to Palestine on the *Exodus* ship. As I was growing up, perhaps due to some strange reaction to my parents' survival against all odds, I developed a weird love for danger. By the time I was fourteen, my kayak filled with water and sank in a storm quite far from land. By the time I was sixteen, I was diving with sharks in the southern tip of the Sinai desert. By the time I was seventeen, my love for speed resulted in a car literally going over me and my bike. By the time I was eighteen, I was flying jets, and an eagle smashed at high speed into my cockpit window, spraying my face with shattered glass and bone. By the time I was twenty, I ejected below the death graph of my ejection seat, and a small canyon allowed the additional one second necessary for the parachute to open. At twenty-two, my friend died as his jet exploded while flying next to me in a plane that I was supposed to be flying. At twenty-three, I almost didn't make it out of the narrow burrows of a deep unchartered cave, and at twenty-five, I was para-jumping from the cliffs above the Sea of Galilee. One day I crashed my snowboard so badly that the bindings came out of the board. The man at the repair shop said he had never seen anything like it. The list goes on and on. But perhaps my most memorable near-death experience was with Greenpeace, when I joined a team that was about to board and take control of a toxic-waste-dumping ship in deep waters. I volunteered to be the guy they throw in the water just in front of the advancing ship so that the captain sees me and stops the ship. We went several times with our Zodiac boats around the ship to make sure we had his attention. I jumped, but the ship did not stop, and I had to swim for my life to avoid being smashed by its tip or chopped by its propellers.

Thinking back on all these events, it seems evident that Darwinian physical survival was my creed, that the fight against those flat lines on the hospital screens monitoring the brain and the heart was all that mattered. This was my conviction, and I loved the challenge.

But then something happened. I began to suspect that life is more than just the functioning of the physical organs. After all, zombies, which we consider to be in many ways dead, have functioning physical organs as well. I said to myself, "What if those oscillating lines going up and down on the hospital monitors, measuring my heart and my brain, don't mean much? What if my belief that I am alive, based on these hospital monitors, is nothing but fake news?" I had no idea what I actually meant by these conundrums. Still, I started wandering about through the maze of skepticism by asking myself the extreme question, "Am I really alive?"

In fact, many of us ask ourselves this very question in different versions, though perhaps with more easily digestible narratives, such as, "What is the meaning/purpose of my life?" or "In what sense am I alive?" but also with these gentler, more polite versions, it seems none of us is really sure what we mean by it. It appears as if our intuition is telling us that the medical definition, which is based on heart and general brain activity, is too rudimentary. At some point, we may be content to conclude that the mere fact that we are breathing is good enough proof that we are alive, but then we recall that the most primitive worm, fish, or lizard breathes as well, and has a heart that beats, and a suspicion creeps in that our contentment was ill conceived. Perhaps this is the reason that so many novels and movies suggesting that our understanding of the reality of our human life is very partial resonate with us—for example, movies such as *The Matrix* and *The Truman Show*, and more recently, *Don't Worry Darling*. But to be honest, I was not thinking of such imaginary scenarios, in which life is nothing but a fabrication based on simulations of life. I was assuming that the reality we see around us is, in fact, quite real. But within this reality of being born, growing up, having a family and a career, I was still asking, "Am I really alive?"

My curiosity kept on growing. We humans like to applaud ourselves as being endowed with a worthier form of life relative to other animals, but where is the proof? Watching all those zombie movies, which describe entities or animals that are biologically quite similar to us, sometimes also called "imitation man" or "dead man walking,"[*] I started to ask myself if I could find a good definition for the difference between a zombie, which we consider to be dead, or at the very least to represent a lower form of life, and a living human, specifically me.

---

[*]The idea of the zombie as some dead man walking goes back to African and Haitian folklore. In the twentieth century, philosophers started paying attention, turning the idea into a thought experiment examining the very foundation of our species, although a clear-cut definition for what a zombie is was never agreed upon. Some other terms were used as well, like "imitation man." The philosophical zombie was introduced by people such as David Chalmers. This is a thought experiment about a hypothetical being, identical to humans in every way, but devoid of the subjective conscious experience. The zombie will tell you that he feels pain when a needle punctures his skin, but it will actually not feel pain. It just knows that it should feel pain. Philosophers are divided if such a concept is helpful.

So the question "Am I really alive?" could be turned around and asked as, "How do I know I am not a zombie?" Statements like that of philosopher and cognitive scientist Daniel Dennett, who said in 1992, "Are zombies possible? They're not just possible, they're actual. We're all zombies: nobody is conscious," just injected more fuel into the fire that was starting to burn within me.

I thought to myself that if merely breathing, or having a functioning heart and brain, or having metabolism, or an evolving (mutating) DNA, is a good-enough proof of life, then I am alive in the same way a zombie is alive. My understanding grew increasingly stronger that to differentiate between a zombie and human life, I must define human life in a way that goes beyond physical bodily functions. But what is there beyond bodily functions? Let us simply call it, for now, mental levels.

But if indeed a meaningful definition of human life requires going beyond bodily functions, there seem to be dire consequences. Specifically, as there is such a spectrum of human behaviors—perhaps pointing toward the existence of different mental levels—it could not be that all humans are alive in the same way. Just as there is a huge spectrum of mental capacity (and under some definitions even consciousness) between a mold and a worm, and between a worm and a fish, and then between a fish and a cow, and between a cow and a human, it stands to reason that also within the human species itself there would be different levels of consciousness and life, with varying levels going all the way from zombie state to the highest human state. As the brain is so complex—so I argued with myself—there are probably very different levels of brain functionality or mental capacities that give rise to different levels of (mental) life. Obviously I was not thinking about superficial scales such as IQ tests, or anything that could be connected to the trivial classifications of race or gender, as all these have proven themselves to be completely bogus when the philosophy of humans is considered.

I must admit that at first I was deterred from thinking of any kind of classification of human life that might indicate that there are varying levels of life within the human species. History has taught us that dark agents always find a way to use such classifications for their own evil agendas of stripping part of our species of their basic human rights (see cautionary statement*). Still, I

---

*And indeed there is a lot to fear when it comes to classifying the quality and merit of different humans. The Nazis and their evil agenda, and black slavery in America, are the most well-known examples, but we should remember that such ideas are common among many societies and schools of

thought to myself, as long as we commit to never allowing ourselves to go into these very dark corners of our civilization, if a deep dive into who we are can secure the future of our species and perhaps even save it from some doomsday annihilation, we must be in favor of such an endeavor, right? But mostly, when I was young, I believe I was not pondering the future of our species. I think that at the core of it, I was simply curious about who I am.

I quickly discovered that I am not the first to suspect that there is more to us than just the body. For example, the mind-body problem, which asserts that the mind and body are fundamentally different in nature, is an ancient debate. Already Plato and the Buddha thought about the relationship between these two separate entities. So I started to ask myself, what are the things that those existing hospital monitors of general brain activity cannot measure? I sensed that many people actually shared this suspicion of mine, that the present measurement of human life, based on those lines on the monitors, is too rudimentary, even if they are not able to clearly put their finger on it, even to themselves. In addition to the numerous science-fiction stories that I mentioned above, suggesting that our understanding of the reality of our human life is incomplete—not to say outright wrong—there are also all kinds of theories, popular among humans, trying to prove that indeed the mind is different from the body, and consequently life cannot be just about the functioning of the body. For example, many people believe in some sort of life force, sometimes called in alternative medicine "the energy field."

---

thought, for example, the eugenics movement. The term was coined in the United Kingdom and then spread to the United States and other countries. They really believed that for the ideal human, you must improve the genetic makeup of the population. In the United States alone, tens of thousands of people were sterilized by force. Even Plato discussed the need for engineered breeding. It could be that the whole thing started with him. Some Nazis even used this as part of their defense in the Nuremberg trials. In modern times, perhaps the first great massacre based on the belief in racial superiority, also called "the forgotten genocide," happened in Namibia in 1904–1908, at the hands of the soldiers of the German Kaiser Wilhelm II. Another relatively unknown story is what Belgium has done in Africa. For example, they used to measure the locals for the size of their skull, the length of their nose, and the shape of the eyes, and that's how they convinced the Hutu and Tutsi populations that there is indeed a basic difference between them and that one was superior to the other. This started the genocide ball rolling. It took it seventy years to roll, but roll it did. And if that was not enough, the Belgian king, Leopold, got his money from rubber in Congo. Obviously, slaves were used. Millions died by murder or due to conditions. A common practice in those days was to chop off hands of locals if they did not meet the daily quota. These are just a few of an endless list of examples of how dangerous classifying people can be. On the other hand, understanding and defining quality is how our species may advance. It's a fine line to walk, and we should continuously be looking over our shoulder to make sure Satan is not smiling. A good insurance policy is to continuously empower instruments protecting fundamental rights, such as the Universal Declaration of Human Rights. Concerning eugenics, see also footnote in the epilogue.

There are of course all those who believe in the concept of the soul, which is independent of the physical body. Others believe in the field of emergence, which claims that complex systems develop completely novel qualities that cannot be understood by simply considering the sum of the system's parts. So it seemed that quite a few people shared my feeling that we humans need to dig deeper into our core hardware and software and, if possible, invent new ways to measure life. So how can one answer the question "Am I really alive?"

I began to read everything I could, and I came to the conclusion that putting aside metaphysical concepts, such as the soul and the life force, science would eventually solve this enigma of how to measure human life through the study of consciousness, as consciousness seems to be what is unique about the human species relative to all other animals, and probably also what differentiates us from zombies.

It seemed clear to me that scientists would eventually find out exactly what consciousness is. After all, at the disposal of science were great clues regarding consciousness, such as the fact that a wide range of chemicals with anesthetic capabilities can turn it off, and that some chemicals, such as psychedelic drugs, can expand or even alter the state of consciousness, or that a big part of the brain, the cerebellum, has very little conscious activity, not to mention that our consciousness is completely modified during sleep. Furthermore, scientists had all these new brain probes such as the EEG, MEG, and MRI.

But the field of consciousness did not seem to be able to take off, or at least show some signs of converging. In 1989, when I was twenty-six, Stuart Sutherland, a British psychologist, said, "Consciousness is a fascinating but elusive phenomenon; it is impossible to specify what it is, what it does, or why it evolved. Nothing worth reading has been written on it."[1]

However, as described in this book, since then an explosion of innovative theories, such as integrated information theory (IIT) and global neural workspace theory (GNW), as well as impressive empirical studies (e.g., with the above probes), have emerged.

I wanted to delve deeper and deeper into these theories, because I felt that I could not live without knowing who or what I am. One may even say I became obsessed with it. Am I just my body? Am I a zombie? Am I actually alive? According to which definition?

Alas, eventually life distracted me from questioning life. Life is so very good at creating distractions, isn't it? The job, the family, love, money, hobbies like the guitar, global problems like the environment and the extinction of wildlife, human rights, and of course my passion for physics—they all swept me away. The whirlpool of existential emotions and thought that I was previously drawn into had subsided.

But then one day it came back in a big way, like some dam that had been waiting to break. As I explain in the introduction, serendipity was responsible for that first crack. Serendipity and a smart New York City filmmaker named Eve.

2. ARTIFICIAL INTELLIGENCE

But before I tell you about Eve, I would like to briefly present the main topic of this book, that is, AI and the mind-boggling introspection experiment it is about to enable.

We all know that the all-powerful AI is coming. When? Soon. It's on everyone's mind. Some, in a somewhat agnostic manner, say it is poised to become the highest-impact event of the twenty-first century. Others, taking a more positive stance, say it will create a beneficial leap analogous to that of the industrial revolution.

On the other hand, there is trepidation of imminent danger. Some fear that this future of superintelligent machines would leave humans lacking careers, income, or purpose. There are those who dread the ability of omnipotent AI to create fake news. Others talk of an out-of-control arms race between AIs serving different powers, while some go as far as prophesying the complete annihilation of the human race at the hands of these machines—the new kings of the planet.

The bulk of the horror comes from the fact that AI may become conscious. This anxiety, justified in my opinion, has now reached a boiling point. The striking coming-of-age of AI (e.g., the winning of the first prize in an art competition by the program Midjourney,[2] or the fact that AI may eventually be externally undifferentiable from humans,[3] so-called android or humanoid) does little to alleviate the fear. Indeed, Ray Kurzweil in his book *The Singularity Is Near* predicts an exponential increase in the relevant technologies. This imminent conflict has already found its way to court with lawsuits for infringement of copyrights by AI.[4] But this is only the tip of the iceberg. In an open letter initiated by the Future of Life Institute, a pause in AI development was called for.[5] Similarly, a Microsoft team contended that GPT-like programs show "sparks of artificial general intelligence."[6] To fight off this onset of terror, AI developers have been scrambling to find solutions, and these include what has been termed "AI human value alignment," aligning AI with the social, ethical, and moral norms of humans.

An endless body of books and articles have been written about AI by brilliant people, and this begs the question whether there is anything completely new left to be said. As explained step-by-step in *The Human Test*, the answer is a resounding yes! Even if these machines do not achieve consciousness within our lifetime, their impact on us will be life transforming in ways much more profound than those we have so far imagined, and this impact I speak of has been missing from the AI discourse. It has all been percolating under the radar, and the random dots are yet to be put together coherently.

Specifically, nonconscious AI (hereafter simply AI), together with Big Data, are already building an individual model of each of us through our digital footprint. For example, social media (e.g., Facebook [now Meta], Instagram, Twitter [now X], TikTok, Netflix, Spotify, YouTube, and even news organizations) are becoming extremely proficient in collecting data and predicting what content you will engage with. For the most part, algorithms deciding which posts will pop up on your screen have been optimized for that. These algorithms can see exactly what content you spend time on or react to (with likes, comments, resharing, or just strolling through and dwelling on some content for some time). Even you, the user, cannot describe your behavior as well as they can. This information is fed into deep-learning algorithms that use a huge amount of data with billions of data points to decide what to show you next. These powerful systems are well hidden behind unintimidating names like "recommender systems" and "user preferences." This trend will quickly spread to all walks of life. It's all about designing luring traps, customized and optimized to each person, so that you end up acquiring whatever product or service is being offered, whether materialistic, economic, social, spiritual, or political. But to do all this efficiently, they need to know in advance what your choices are going to be; in other words, they need to predict you, and me, and every one of us. So at the end of the day, it's all about human predictability. The more predictable we are, the more successful these algorithms will be.

Where will the information used by these algorithms come from? Data-harvesting sensors will be embedded in everything you touch, use, or interact with (many of them under innocent-sounding, not to say alluring, titles such as "smart home," "smart car," or "smart clothes and accessories"). Our digital footprint has already been coined our individual "intelligent history." Perhaps the most in-depth data harvesting will eventually be done by the gaming industry (at the time of the writing of this book, as many as 3.2 billion people were playing video games); to become more and more attractive, they will need to study each and every one of us in great detail, revealing our deepest driving mechanisms for the purpose of predicting our choices. Based on endless data harvesting of our every move—a trend that will continue to grow exponentially—and furthermore utilizing computing power beyond what we can now even imagine, AI will be able to predict what we want for breakfast, what shoes we will buy, what political party we will vote for, and what partner we will fancy in a bar. Indeed, these deep-learning machines will know us better than we know ourselves.

Imagine yourself in your car at an intersection. You have a red light, and you know that as long as the light for the crossing traffic is still green, there is no way your light will change to green. When the crossing traffic light turns red, you can predict that in a couple of seconds your light will be green. You are able to make this prediction because you know that behind the traffic light there is a rudimentary machine, a deterministic machine, where *deterministic* means that situation A always leads to situation B. As explained later in this book, this has been termed the Laplace demon. While sitting in our car at the junction, we do not reflect on the significance of this predictability, as we are used to it. But what if AI and Big Data prove that we humans are predictable in the same way? The paradigm-shifting ramifications are detailed throughout the book.

For example, a topic that is analyzed in depth in this book is that of *free will*. There have been countless scientific arguments for and against the existence of free will in humans. The recent book *Determined: A Science of Life without Free Will*, by Stanford behaviorist Robert M. Sapolsky, is the latest in a mesmerizing myriad of opinions and arguments going back centuries. However, clear-cut and decisive scientific proof is yet to be given. Indeed, quite a few scientists still believe in free will. As predictability is incompatible with free will, the predictability experiment described in this book will indeed be able to put this debate to an explosive rest.

To conclude this preface, we may say that *The Human Test* in fact begins with the last sentence of Yuval Noah Harari's *Homo Deus: A Brief History of Tomorrow,* in which he asks the reader, "What will happen to society, politics and daily life when non-conscious but highly intelligent algorithms know us better than we know ourselves?" This book now in your hands delivers a highly focused and radically new answer orbiting around human predictability.

But let us first go back to that day that changed everything, the day I met Eve.

Introduction

1. EVE

If the preface sounds too much like the stuff of science-fiction dystopias, let us now start our voyage toward the concrete.

In my quantum science laboratory, my students and I hold single atoms undisturbed inside a vacuum chamber, where conditions are similar to those in deep space and, with the help of lasers, we communicate with them via windows in the metal chamber. In fact, we communicate with the atoms much like humans communicate between one another, by physically perturbing the space between us. We send atoms information by modulating the laser light that we shine on them, just like humans modulate air density to create sound waves traversing between one human and another. The atoms do something with the information and send us back different information, again by modulated light. For example, if, with the help of the laser light we shine on them, we ask them to measure time (atomic clock), calculate something (quantum computer), or even search for dark matter, they will send us the answer through the light they shine back at us. As light is nothing but electromagnetic waves, we are actually sending information to the atoms just like our local radio station sends audio information to the radio in our car. As my communication with the atoms became better and better, I could propel myself into their universe, turning them into my teachers, learning from them how their quantum world works. I was interested in understanding the roots of this fabulous theory, which is one of the two pillars of modern physics, the other being the theory of relativity. My interest in the measurement and definition of human life was all but forgotten.

And here is how serendipity struck. One day, a very curious and insightful filmmaker from New York City named Eve visited our lab. Now that I think back, I'm not sure how she got there. I seem to recall that she visited our university as part of some group and, when she had some free time, she asked to see an interesting lab and they sent her to me. I remember that I was busy that day, so I took her to the lab and asked the students to show her around. Before I left, I spent some time with the group to make sure she was well taken care of. There was something a little weird about her. She was not just very curious and insightful, but she also smiled a bit too much. It was not out of politeness or some empathy toward the hardworking students. This became evident as at times her smiles turned into short inexplicable bursts of laughter. She could see that the people around her felt it was awkward, and she clearly tried to mitigate these bursts. For the short time I was there I managed to correlate between her bursts of almost hidden laughter and what was being said in the room, and I came to the conclusion that she was laughing every time she learned something utterly new, some astonishing fact. At first I felt that I was simply observing the joy of her intellect, or the delight of her sense of beauty, but then I thought it must be more than that. It was as if she was laughing at how ridiculous things seem to be in nature, how improbable. I left the lab. As I was walking to my next meeting, I recall wondering if her name, that of the very first woman, had anything to do with it, as if she had convinced herself that within her lay something of the original Eve who, after millions of years, had awoken to observe what humans have become and what they have learned.

In those few hours, Eve changed my life, because she made me completely undo and redo my perspective on life, and this is how: After a few hours in the lab, mostly with my students, I guess she could see how we communicated with the atoms, and then, when I arrived to take her back to her university hosts, she turned to me and I could see that she was not wearing her smiling face anymore. Her serious, almost somber face seemed to be some kind of overture to what she was about to say. She looked straight into my eyes and asked, "Are they alive?"

I was shocked. I felt blindsided. Her question exploded in my head like a huge tidal wave hitting a humongous drum. With those three words, she brought back my obsession with the definition of human life and, with it, an endless stream of thoughts. It took me a few seconds to regroup. I told myself that since she was merely talking about atoms, the answer must be simple. I instinctively reflected on NASA's effort to define life as it searches for signs of it throughout the cosmos, using criteria such as *self-reproduction with variations,* or *a self-sustained chemical system capable of undergoing Darwinian evolution,* but I quickly felt that these criteria fall short of appreciating the significance of the role of communication, such as the one taking shape with the atoms. When our own brain does this kind of change of information between what it receives and what it emits, we say that it analyzes the information, namely that it thinks, and if it thinks, then it must be alive. (This is probably what René Descartes meant with his famous declaration, "I think, therefore I am.") But what is the real difference between this activity of our brain and what the single atom in my lab does when it communicates with us?

I now recall that, looking at her, seconds after she asked the question, I briefly thought of Confucius, who said that if you are the smartest guy in the room, you are in the wrong room, and I was thankful she was there.

I then remembered a brilliant man, a good friend of mine, Claude Cohen-Tannoudji, who received a Nobel Prize for the exact description of how these atoms interact with light. Claude described them as a machine, a very convoluted machine, but just a machine. This means that a mathematical formula could predict exactly how an atom would behave under certain conditions, for example, when it is hit by a photon. Still, standing there in the laboratory after Eve asked her question, I sensed a sort of dryness in my mouth, as I usually do when I feel I am facing my ignorance.

I told her I'd have to think about it. It then dawned on me that before I could answer her question, we would have to agree on some terminology. One of the three words in her question was the word *alive,* so to answer her question we have to first agree what is meant by this word. As we started walking away from the lab, I said, "What do you mean by *alive*? Do you mean what we think of when we talk of human life?" She paused her walking, and I echoed her pause. "Yes, I think that's what I mean." "So, with your permission I will first try and address the question of human life, I mean, in what way is a human alive, OK?" She nodded in agreement. We continued walking. It seems my question about the definition of human life has come back to haunt me, I said to myself. I was also reflecting that the two questions may be identical, as I was not exactly sure if there is any difference between that atom in the vacuum chamber and me, for I am my brain, and my brain is nothing but atoms.*

I brought her back to her group. A minute or two earlier, she made a rather dramatic request. I will tell you about it toward the end of this introduction. "Be well," I said as we parted. As I was walking away, I was surprised to sense that my words were coated with that mixture of warmth and sadness that you get when you part from someone who has managed to find their way into the deep hidden corners of your heart and mind. While "I am not sure we are alive" may not be the greatest pickup line in a bar, it seemed that both of us were obsessed over it, and this spontaneously created a bond.

Later on in my office I could not focus on the work I was supposed to do. I was busy contemplating that so many people are using the words *birth* and *life* and *death* as metaphors rather than as an accurate physical description of what life is. For example, I remembered once reading the words of a socialist, Moshe Beilinson, who said something like, "We set out to change the world . . . to end the enslavement of people by people. When were we born? It was when the first spark of freedom and self-esteem lit in the heart of the first slave." He used the word "born" to describe when life began, but clearly as a metaphor. However, this filmmaker visiting our lab was not interested in a discussion about metaphors. Like myself in previous years, it was clear she wanted precision; she wanted a scientific definition.

Eventually I did send her an answer by email, but I was dissatisfied. She was dissatisfied as well. She was polite, but I could feel her discontent lurking

---

*To clarify, in this book we focus on a pragmatic measurable definition of (mental) life within the human species. We are not addressing the question of how to arrive at a universal definition of all life, an important effort in its own right, undertaken, for example, by NASA. See, for example, a popular article by Carl Zimmer (2021) in *Quanta* magazine: "What Is Life? Its Vast Diversity Defies Easy Definition." See also the book by Carl Zimmer (2021), *Life's Edge*.

between the lines of her reply. It was evident she did not want to beat around the bush. She wanted a razor-sharp, to-the-point answer.

Like a typical academic circling his wagons, my email to her started with a look at the history of the ancient debate over the mind-body problem, which asserts that the mind and body are fundamentally different in nature. As noted, Plato and the Buddha, among others, already thought about the relationship between these two separate entities. An endless number of philosophical masterpieces, as well as poems, prose, and plays, have been written about it. I also told her about the field of emergence, which claims that complex systems develop completely novel qualities that cannot be understood by simply considering the sum of the system's parts. For example, some claim we cannot predict the shape of snowflakes or fractal water crystals forming on glass from the knowledge we have about water molecules and, similarly, that we cannot predict the exact ripples seen in sandy surfaces hit by winds or currents by knowing the atomic composition of sand. In the same way, although it is built from atoms, it could be that the brain has features that are fundamentally different from those of atoms, and in this sense, it may be that even if we eventually think the atom is dead, the brain (and the human behind it) may still be alive, and, vice versa, if we understand in what sense a human is alive, it could be that the single atom does not meet the human criterion for life. I concluded my email to her by saying that her question reminded me of the words written in Yiddish by the poet and playwright Abraham Goldfaden in the nineteenth century: "They say that I am not who I am—and it scares me, for if I am not who I am—who am I?" I was blowing smoke, and I recall I even felt embarrassment over my near-random contemplations. I did not have what she wanted.

In the many months that followed, every day as I watched the atoms in my lab, it became clear that many layers of her question remained hidden from me, and I was adamant not to send her another answer before I felt I had reached the very essence of her question. And so, this book, *The Human Test*, my answer to her, started a life of its own.

As time passed, something strange happened. For some reason I felt that from our discussions over the course of that day, and from the few emails she sent me afterward, I already knew her quite well, and I started to imagine what questions she might still have. She became the dissenting voice within my mind, an avatar roaming on a big Harley Davidson motorbike inside my skull, emitting, from time to time, bursts of laughter. It was clear she was extremely smart, and I had to keep up. In this introduction I outline some brief examples of the sort of discussions I had with her.

Again and again, I asked myself if my answer to Eve was now ready. Time and again I could not shake the suspicion that it was still not complete. It eventually took a very long time to hatch from its cocoon.

2. GOING BEYOND CONSCIOUSNESS

The first thing I wanted to do was to go back to my old interest in the field of consciousness. I again reaffirmed my belief that, through the study of consciousness, science would eventually be able to define what is special about our brain and consequently to define human life, and so it was that the meaning of consciousness became for me the key to differentiating us from zombies, or from atoms for that matter.

But then, after refamiliarizing myself with the latest scientific work, I decided I must go beyond consciousness, as it was still so horribly ill defined, both theoretically and experimentally. This is evident in the many diverging opinions regarding consciousness that continue to linger to this day in the scientific community. The more I read, the more it became clear that the study of consciousness features an extremely high potential to eventually provide us with a deep understanding and should, therefore, be continued, but the chances of it converging any time soon to some clear-cut insight, not to say definition, of (mental) human life, or what it means to be human in terms of higher brain functions, are slim.

In fact, I must say that the more I read about the present theoretical and experimental search for consciousness, the more it resembled the search for dark matter (a type of matter that we physicists are working hard to identify): In both cases, we are quite sure it's there, although currently we have no idea what it is, but we are developing lots of interesting theories and experimental probes to eventually understand it. It is crucial that we continue to study dark matter just as it is of paramount importance that we continue to study consciousness. However, in both cases, it may take a very long time.*

I therefore wanted to look for something simpler that could provide some operational definition of human life within our lifetime, something that exist-

---

*Nevertheless, my brain surprised me with its optimism, when it convinced me to build an atomic dark matter detector in our lab, a detector that is still working and searching to this day.

ing, or near-future, technology could actually measure with crystal-clear verdicts. When I say *measure*, I mean that some instrument can assign a value, a number, like we do in experimental physics for mass, or velocity, or charge. In physics, when we are able to measure something, we call it an "observable." Hence, I wanted some realistic, tangible observable for human life. At first I was very skeptical about the possibility, but I said to myself that if I am to deliver a meaningful answer to the NYC filmmaker, I must try and go beyond the vagueness that currently dominates the discourse.

I also defined for myself a second, related goal, which was to try and find an observable that quantifies human life in a way that is determined by the outcomes of the brain, namely, the actions of a person. Actions are a clear-cut data point, whereas internal brain processes, such as those measured in the studies of consciousness, are much more ambiguous. Of course, I was well aware that humans have been characterizing human behavior, namely brain outcomes, for thousands of years. That's what we call good and bad, smart and stupid, warm and cold, spiritual and rational, ethical and immoral, flamboyant and reserved, and so on. But all these lack straightforward scientific definitions; they are subjective. I wanted an objective observable for which I could get a clear-cut number. Again, at first, I was skeptical about the possibility.

As described throughout this book, as I was peeling back layer after layer surrounding Eve's question, the predictability of our brain and its decisions (namely, brain outcomes), and thus the predictability of us humans, slowly revealed itself to be the observable I was searching for. It also seemed that AI and Big Data would be able to measure this observable. But what does predictability actually tell us about us humans? And can AI and Big Data really predict us? At the beginning there were more questions than answers regarding this surprising observable.

One thing was becoming evident quite quickly. As I was diving deeper and deeper into present-day theories of consciousness in search of an answer, an important insight was emerging, a logical connection which, it seems, only a few have made: The idea that the brain and its outcomes are in principle predictable is, in fact, currently being established inadvertently by the different theories that try to provide a mathematically accurate description of the brain and of consciousness. Obviously, every theory that has the eventual goal of describing the activity of the brain by some mathematical function inadvertently supports the notion of a deterministic brain, as mathematics is deterministic! (as long as there is no stochastic, i.e., random, term built into the mathematical equation). For example, one of the popular theories of consciousness in our time, integrated information theory (IIT), attempts to provide a numerical value for consciousness which is coined *phi*, and which aims to represent both subjective experience as well as the empirical evidence. The next logical step is to realize that anything that can be described with mathematical accuracy is eventually a predictable system, so talking about predictable brains or predictable humans is indeed a rational leap. Of course, even a mathematical formulation can only make accurate predictions depending on its granularity (i.e., level of detail) and the quality of its input, but in principle, if you believe in a mathematical description of the brain, the concept of predictability should not be foreign to you.

With the passing of time, my communication with the atoms became more sophisticated, and I could delve further into the implications of this new way of looking at the mathematics of the brain.

As a picture started to emerge, I gradually became quite stunned to realize that with the advent of AI and Big Data, an actual experiment that could shed new light on the question of human life, and in fact deliver profound new insight, is already underway, and amazingly no one talks about it.

3. MACHINE DEATH

As my answer to Eve's brilliant question took on a life of its own, it became apparent that the observable that may shed new light on the question of human life is predictability. The experiment measuring the predictability of humans is described in Part I of this book, as well as briefly below. Although I analyze the meaning of predictability throughout the book, here, in a nutshell, I touch upon the immense philosophical impact of this observable.

Rather than speak of zombies, in my discussions with Eve I began to talk of machine death, namely humans dying not biologically but mentally, because they are a machine. I wanted to discuss machine death as it is easier to define what a machine is than to define what humans are, or zombies for that matter. In fact, if by machine we are referring to classical, nonintelligent machines, a good machine may be defined by just one word: *predictability*. I have already mentioned in the preface the predictable behavior of the traffic lights at a road junction. In a more complex example, if a good car engine gets the proper amount of air, water, gasoline, and oil, it is completely predictable; you know exactly where the pistons will be in one hour or one day. If machines were not predictable, we would not have made it to the moon and to Mars. The *Voyager* would not still be sending us data from the edges of the solar system. It is this concept of predictability that I focus on, because it has profound implications for our understanding of our human life, and in fact it holds the key to answering Eve's convoluted question.

As predictability is the hallmark of a machine, and if we agree that a machine is the opposite of *human life*, it stands to reason that being a predictable human is a sufficient condition for what may be called *human death*, where by death I obviously mean mental death and not biological death. What if a disruptive new technology, now making its debut, manages to prove that we, namely our desires and choices, are predictable? It is not the technology that will make us predictable. If we are predictable, we are predictable by design; namely, this is how nature made us. The new technology will simply prove that this is our reality.

The first thing to note in the interplay between predictability and human life is that if an individual and his thoughts may be predicted by some external agent, that individual cannot have *free will*. So, while it is hard to define exactly what free will is, it is easy to define its opposite. This is the first of many dramatic consequences.

Some scholars claim that AI, with all its superintelligence, is going to be the highest-impact event that will take place this century. Here, I beg to differ; while it is evident that superintelligent AI is a game changer, the new understanding of how predictable we are will give rise to an even higher-impact turning point in the story of the evolution of the human species as, in many ways, it heralds our extinction as a free species. *Extinction* is a big word, so what exactly do I mean by that? Specifically, as predictability is incompatible with free will, an experiment showing that humans are predictable will prove once and for all that there is no free will; namely, considering the most fundamental meaning of *free*, we can no longer be defined as a free species. Some may say that free will is not that important, but if some external entity can know what we will decide before we decide it, what is there really left of the promise, to say nothing of the beauty, of our species? If we are predictable, we are really nothing more than the sum of our predictable single atoms.

Still, superintelligent AI and Big Data are indeed game changers in this story of our possible extinction as a free species, as it is the proof they will provide about us being predictable that will transform us from a race of humans to a race of machines. Based on endless data harvesting of our every move and utilizing computing power that is beyond what we can now fathom, AI will know us better than we know ourselves and will be able to predict our every choice, or will it?

One day, I was drinking my morning university coffee, and just as I was thinking that this defines the opposite of exquisite coffee, Eve's voice, in what has now become a customary routine, began its almost daily roaming around inside my head. She asked me, "Beyond the way nature made us, are there humans promoting human predictability?" "What do you mean?" I replied. "I mean, are there forces in our society that would be happy to see us being predictable and are already pursuing the measurement of our predictability?"

I thought about it for a few days, and I told the Eve that was inside my head the following: At present, it is primarily capitalism and the business sector that are pushing forward the measurement and utilization of human predictability. The potential economic gains are humongous, and as I have already noted, the ever-growing capabilities, which are currently well hidden behind innocuous names like "recommender systems" and "user preferences," are intended to design traps, ploys, customized and optimized to each individual person based on the ability to predict the choices of that person, so that consumers are lured into purchasing whatever product or service a company is offering.

Human predictability is also promoted indirectly, at times brilliantly, by those who claim that we need to take measures to reduce variability in human decision making, as such variability is tantamount to lapses of judgment. For example, Nobel laureate Daniel Kahneman and his colleagues claim exactly that.[1] While variability is indeed a cause for lack of accuracy and efficiency and takes a negative toll in various human endeavors such as medical treatment, the justice system, and customer service, no one has considered the price humans will be required to pay when the minimization of variability reaches a level where it will increase predictability.

The political sector is not far behind the business sector in driving the field of predictability, just as it was quick to capitalize on the expertise of public relations and advertising firms at the beginning of the twentieth century.

Academic institutions are of course also very curious regarding any new understanding of how we are built. In addition, as the investment of industry in academia increases, the pure curiosity of academics is unfortunately becoming more and more aligned with the needs of the business sector. Indeed, there are many forces at work aiming to prove and take advantage of the predictability of humans.

She then asked, "If we are proven to be predictable, will our day-to-day life be any different from what it is today?" I answered: Not at all. We will still go to work, still have children, and still have loving relationships and business ventures. If anything, we will be happier, as those who are predicting us will be better able to understand our needs and address them. "Why would they want to supply our needs?" she asked. Well, for the simple reason that happy people ask far fewer questions and are easier to control.

"So you are saying that mentally dead people could be happy and enjoy well-being?" she reflected. I believe so, I answered, adding that happiness and well-being are subjective experiences that could be felt even if you have no free will. On the other hand, I noted, the lack of free will that comes with measured predictability is an objective statement, a paradigm shift in the understanding of who we are.

But then I stopped talking. A dark cloud was suddenly looming above me. I whispered some words to myself. "Have you finished with your unintelligible soliloquy?" she suddenly uttered. I tried to raise my voice: We may not even know that we have become predictable. She made a face. It was not a laughing face. On the contrary, it was the same one she had made in the lab quite some time ago when she asked me if the atoms are alive. And now it was clear she wanted to know more. I said: Commercial companies and social media, homeland security institutions and rulers, may not be in a great hurry to reveal their cards, as they hope this technology will provide them with an economic, security, or political edge. Following the Facebook/Cambridge Analytica data scandal regarding the use of harvested data for political advertising in the Trump and Cruz 2016 campaigns, future users of such data will surely become more cautious about unveiling their moves. On the other hand, humans at large will perhaps not be so anxious to acknowledge the new trend, as they will not want to be forced to face the philosophical consequences. Even if a certain individual is not disposed to think philosophically, the ego simply cannot accept such a demotion, as all of our egos like to adorn themselves with the crown of a high-level being, which typically means believing that we have free will and, as noted, predictability is the opposite of free will. Such a demotion would be like asking the king of England to start doing the dishes in the local pub. So it seems that both sides of the equation, those who do the predicting and those who are the predicted, have strong incentives to keep this whole topic, which may be rightfully named the most profound experiment ever done on humans, under the radar.

She replied, "Could we be living, now or in the near future, in a social pyramid of predictability, where each layer of society can predict the layer beneath it, as each layer has more harvesting and predicting technology than the one below it? But who will be at the very top? A dictator, a computer, some alien ruler?" I had no answer.

Walking home that day I briefly pondered: In case Earth is ever attacked by sinister aliens, if humans had the ability to become less predictable, it would be much harder for the aliens to fight humans and conquer Earth. In such a case, fighting predictability would even ensure biological (i.e., physical) human life, namely life as we humans have defined it thus far.

But this was a one-time reflection. In my answer to Eve, I was not thinking of science fiction. Far from it. Her demand for accuracy required that I analyze and extrapolate the very much nonfictitious facts of the rising technology and our brain, and in what way this differs from what we know about simple single atoms.

4. A CONNECTION TO CONSCIOUSNESS

One afternoon, her apparition and I were sitting together on a bench on campus, and she asked me whether there was a connection between predictability and consciousness. Her question was extremely relevant. Indeed, when thinking of a new parameter to characterize a system, a new observable, it is always prudent and insightful to view it also in the context of the traditional observable that is already established in the thoughts, language, culture, and experiments of humans. Hence, if I am to seriously consider predictability as a powerful new observable that can shed new light on who we are within our lifetime, I must also wonder about any possible connection to consciousness.

So, what may be the connection, or even correlation, between predictability and consciousness? I asked myself. As a first illustration, I started thinking together with the Eve roaming around inside my skull of the detailed knowledge that has been accumulated regarding mind-altering drugs. I believe many cognitive scientists would agree that mind-altering drugs induce a higher or expanded level of consciousness. For example, Anil Seth, a cognitive and computational neuroscientist from the United Kingdom, states, "I learn to my surprise that hallucinogens really do take you to a higher level of consciousness."[2] On the other hand, many would also agree that mind-altering drugs induce divergent thought or associative thinking, namely, creativity, which seems to be tantamount to less predictability.* It is not by chance that so many artists find it necessary to use hallucinogens.[3]

Specifically, painters and sculptors, musicians and dancers, poets and authors, for many years have all recognized that mind-altering drugs are a gateway to creativity. For example, it is well known that Pablo Picasso smoked opium. He even stated that the smell of the thick opium smoke was "the most intelligent smell in the world." It is also well known that in music, Bob Dylan, the Beatles, Jimi Hendrix, and many other famous artists used drugs, specifically LSD.

Another person famous for his creativity, Steve Jobs, stated that LSD helped. Jobs was actually at some point in touch with Swiss chemist Albert Hofmann who, in 1943, accidentally discovered its effects on humans.† This vast history of artists and creative technologists finding mind-altering drugs advantageous would suggest that mind-altering drugs are the opposite of predictability, or in more popular terms, the opposite of more of the same. (It is

---

*Indeed, after writing the core of this book, I found that this connection between creativity and less predictability is being contemplated by a few neuroscientists as well. For example, in his 2024 article "The Neuronal Basis for Human Creativity" (*Frontiers in Human Neuroscience* 18:1367922, https://doi.org/10.3389/fnhum.2024.1367922), neuroscientist Rafael Malach argues that "recent experimental and modeling advances in our understanding of the spontaneous fluctuations offer an explanation for the diversity and innovative nature of creativity, which is derived from a unique integration of random, neuronal noise on the one hand with individually specified, deterministic information acquired through learning, expertise training, and hereditary traits." Obviously randomness is tantamount to less predictability, and hence his connection between randomness and creativity is analogous to the connection analyzed in this book between creativity and less predictability.

†Jobs said: "LSD shows you that there's another side to the coin, and you can't remember it when it wears off, but you know it. It reinforced my sense of what was important—creating great things instead of making money, putting things back into the stream of history and of human consciousness as much as I could." https://money.cnn.com/2015/01/25/technology/kottke-lsd-steve-jobs; https://www.businessinsider.com/steve-jobs-lsd-meditation-zen-quest-2015-1.

perhaps interesting to also note that there are recurring claims that the leaps of thought that gave birth to modern civilization some two thousand years ago were connected to the use of mind-altering drugs.[4])

It is therefore quite evident that our new observable may be somehow connected to what we call consciousness, as both are affected by mind-altering drugs, and furthermore they seem to adhere to an inverse relation, namely, that a higher consciousness is connected to less predictability.

Of course, in order for a correlation to be established, one needs to show that a change in predictability is always accompanied by a change in the level of consciousness. This requires enormous scientific work that is beyond the scope of my discussion with Eve. Furthermore, correlation is not causation, and finding a causal relation between the two is an even harder undertaking. In addition, creativity itself has never been clearly defined or quantified. An endless number of research studies and books have been written on the subject, for example, addressing the fascinating question of whether AI can be creative[5] and why and how the AI program Midjourney won first prize in a 2022 art competition.[6] Nevertheless, solid hints are good enough for our purpose, and we continued to contemplate this rather bizarre and fascinating connection.

Going back to Anil Seth's statement that hallucinogens take you to a higher level of consciousness, while it is clear what is meant when we speak of a change in the degree of predictability, it is much less clear what is meant by a "higher level of consciousness," especially if it is within the human species. I am of course not considering here trivial scales of consciousness, examining basic functionality of the brain after an injury as is done with the *Glasgow Coma Scale* (GCS) used in hospitals. I am also not speaking of other less trivial, yet still quite rudimentary, altered states of consciousness like hypnosis and meditation.[7] Nevertheless, these hint at the fact, or even imply, that speaking of varying levels of consciousness stands to reason, even within the class of physiologically healthy brains.

So what does this "higher" mean scientifically? At this point, not very much. But first hints are coming in. For example, using brain-imaging MEG technology, researchers measured the tiny magnetic fields produced in the brain and found that, across three psychedelic drugs, one measure of conscious level—neural signal diversity—was reliably higher.[8]

It seems that the intuition of cognitive scientists is that in the future, different physiologically healthy brains will be found to exhibit varying levels of consciousness. Indeed, as noted, present-day theories such as IIT are laying the groundwork for a mathematical evaluation of the level of consciousness in individual brains. As consciousness and predictability are most probably connected, one may conjecture that similarly, in the near future, tests of predictability will find that different people have different levels of predictability. In fact, as predictability will most probably be measured in a reliable way long before consciousness, the experimental uncovering of varying levels of predictability may constitute an important prelude to a later discovery regarding varying levels of consciousness in healthy brains.

But even without the above-hypothesized connection between consciousness and predictability, the important bottom line is that different values for our observable, predictability, are related to different levels of creativity, and as there is quite a consensus that different brains possess different levels of creativity, it stands to reason that different brains have different levels of predictability as well.

Eve and I were now silent, contemplating this new conclusion that different people will probably be found to have different levels of predictability. On the campus bench, as people were passing by, we pushed our heads back, looking at the sky, wondering what level of predictability we ourselves possess.

5. THE PILL

Amazingly, evolution may have developed some kind of mechanism to fight against predictability, not because of some philosophical aesthetic but because of the enhanced survival probability one has if they not only have the ability to *exploit* a situation (which may be done by some predictable deterministic feedback loop) but also the ability to *explore* for new situations (which requires some amount of random decision making, whereby the latter cannot be predicted). Exploring, of course, is tantamount to what we call creativity or divergent thought. However, from the early success of the new AI-based predictive technologies, it seems that for many people this mechanism to fight against predictability is a dwindling resource. As described in *The Human Test*, one may view these two very different functions of the brain through a simplified two-layer model.

One night, as I was lying awake in bed gazing at the ceiling, Eve asked from the depths of my fuzzy mind, "Is there a cure?" Again, her words shocked me. She asked the question I was afraid to ask. If in the lab she asked if the atoms were alive, she was now asking whether there is any hope for humans. For if there is no cure and we are highly predictable to begin with, all is lost.

Thinking of the grim possibility that many of us, if not all of us, are to a high degree predictable, I was wondering whether there is any point in us shouting, "To the battle stations!" This call to arms has saved us humans so many times before. Alas, the atomic bomb that now threatens to annihilate us as a free species does not originate from some crazy North Korean dictator or perhaps some terrorist out to inflict suffering and mayhem, or even from some alien out to take control of Earth. The bomb is already inside of us. The machine virus is built in, a virus within our core software. It is part of our DNA. This is to say that if we are predictable, if we have no free will, it is because that is how nature made us, the hardware and software from which we have been constructed. In fact, it could be that so far this machine virus has been of paramount importance to our survival, but now, if we are to transcend, it may very well have become a horrible liability.

But perhaps there is a pill. A pill that could fight predictability. A pill that would ensure a free mind and the divergent thought and creativity that come with it. It's a long shot. The chances are really minute, but what if? What would be the ingredients of this pill, who would make it, and who would give it to us? The end game of this book, my answer to Eve, is all about the recipe for this pill.

I thought about it for quite a long time, oscillating between optimism and pessimism, until I eventually forged my opinion: as it is believed that creativity may be enhanced (by drugs or by other less dangerous means), we may conjecture that the predictability of an individual brain may be altered as well.

As detailed in this book, I expect there will be two kinds of pills, one available now and the other farther into the future. The first requires that each individual take whatever measures they currently have at their disposal. At present, no one aside from you is going to produce the pill and administer it to you, as these measures are simply not available as a commercial or medical product; and in any case, those predicting you from above would try and stop the mass production of a successful pill any way they can. Currently, only you can make the pill, and only you can decide to take it. The second kind of pill will be made available as a commercial product once the masses demand it and, consequently, once companies understand that there is a huge market for it. Namely, when the time comes, and I expect it will come in the very near future, that our level of predictability may be measured upon our request, I also expect that not too many years later, a person will be able to ask for some external stimulus, chemical, electrical, or even surgical intervention that will lessen his or her predictability. This would be considered some kind of medical treatment. In *The Human Test*, I utilize the above-mentioned two-layer model of the brain to explain how this pill against predictability will work.

This was her final question. *The Human Test* was now ready to be sent to the real her, to Eve.

6. THE BOOK

This book is my answer to Eve, the filmmaker from NYC who visited my lab quite a number of years ago and shocked me when she asked if the single atoms with which I communicate are alive. It took me a very long time to fully understand and appreciate the question. I started out by thinking about the difference between the single atoms in my experiments and our brain, for we are our brain, and our brain is nothing but atoms. Hence, if we are alive, perhaps the atoms are too. But then this thought backfired on me, and I began to think that if atoms are dead, perhaps we are dead too. In my mind, the jury was out. They were not simply out in the next room; they have all run to the underground bunker.

Humans have had great success with the science of life and death, and the result is evident in our ever-increasing longevity. Biological longevity may continue to increase, and as some speculate, our mind may even be awarded eternal life when we are able to upload ourselves to advanced computers. But is that all the science of life and death will achieve? I think not. There are growing indications that science and technology will eventually enable us and indeed force us to completely redefine death, and consequently life. Specifically, if we find that different healthy brains have different levels of consciousness, and if we agree that consciousness is the unique hallmark of human life, we may eventually declare that there are different levels of life and, in the more extreme view of this future, humanity may even declare a certain threshold below which there is no human life, even in what are now considered to be biologically healthy brains. If some form of mental death is much more intricate than the physical death of the brain and the heart—that is, if it may occur even while the biological body is still functioning according to traditional rudimentary criteria—and if this new form of death will define the point at which life really ends, everything will change. Our most cherished convictions will be shattered.

However, if we want some clear-cut answers about who we are mentally within our lifetime, if we want to really understand what is unique in what concerns human life, we must make a bold move. As the study of consciousness is expected to require significant time to converge, we cannot continue to only use the vague terms always called upon in order to define our life, such as *self-awareness* and *consciousness* (see Appendix A regarding the structure of the brain and Appendix B regarding the present status of studies concerning self-awareness and consciousness). Words such as *consciousness* have never had an accurate, quantifiable, measurable definition. Adding additional phrases such as *sentient beings* or *qualia* does not undo this vagueness and ambiguity. As I was trying to complete my answer to the question posed by Eve, it became evident that without a bold move, my final text to her would merely constitute more of the same. My answer to Eve therefore describes the required bold move, in terms of a new experiment and observable, as well as in terms of the interpretation of the results through the lens of a new definition of life.

With this in mind, the book strives for a new paradigm, and it is found in human predictability. The paradigm is simple yet powerful: If you are predictable, you are a machine, and if you are a machine, you are dead or, at the very least, less alive. In popular terms, you are a zombie. You are a complex biological, chemical, physical machine, but a mere machine nonetheless. Human life, which must strive to be the opposite of a machine, cannot be predictable.

When Eve and I became serious about predictability, I cautioned her that professionals from the very wide variety of disciplines studying the complex field of consciousness may feel that the new observable is incomplete, perhaps even simplistic. I recall she took on a serious face, but then she smiled and said, "But such is the nature of a practical compromise." Indeed, she was right; at the end of the day, one must compromise between the wish for a pragmatic operational model that allows us to make simple, clear-cut definitions (in science sometimes called a phenomenological model) and a much deeper, more complex model that is derived from first principles, that is, from the most basic laws of nature. The latter is what is required for a full understanding, while the former is what is needed if we are to get reasonably robust answers within our lifetime. We looked at one another, and it was clear what our choice was.

In summary, predictability has many interesting aspects to it. To begin with, such a well-defined yet, as I found out, insightful parameter allows us to avoid the vague terms predominantly called upon in order to define human life, such as *self-awareness* and *consciousness*. Furthermore, as the single atom seems to be a predictable machine via equations like those written by Claude Cohen-Tannoudji, predictability may be the cornerstone in the search for an objective way to compare between our brain, namely us, and the single atom. Moreover, even in the unlikely event that scientific work with EEG, MEG, and MRI probes brings the entire community of scholars to agree on what consciousness is and how to quantify it, would there be agreement on designating this highly intricate internal mechanistic characterization of the brain as a criterion for human life? Internal signals coming from the brain are so noisy, while predictability, on the other hand, is an outcome of the brain, whereby the data comes from a person's clear-cut actions. The data is consequently much more straightforward and reliable. I thus chose predictability as the objective Rosetta stone around which my answer revolves.

Beyond what appears in the sneak preview provided by the preface and this introduction, my humble answer to the brilliant question put forward by the filmmaker from NYC is divided into five main parts:

First, I begin by treading on solid ground when I describe the profound experiment that is already underway by use of AI and Big Data and which should provide, in the near future, new groundbreaking insight into who we are. While the results of a new experiment are never known ahead of time, preliminary results that are coming in (as described in Part I) allow us to conjecture with a relatively high level of confidence what the eventual results will be.

I review how present-day technology is already well on its way to reliably predict us and how in the near future one may even ask to be tested for their level of predictability, namely, for this new kind of mental death that is being postulated. Here, I also discuss the theoretical basis for the experiment by venturing into a detailed contemplation over determinism, where *determinism* simply means that the future of any physical system is in principle predictable.

In the second part of the book, I present some of the common arguments against the possibility of predictability, such as the soul, free will, and emergence, and I ask whether they are significant.

While this book is mainly about the science of predictability, in the third and fourth parts of the book I engage in a philosophical pause for two different reasons.

In the third part, I take this philosophical pause to acquaint myself and Eve with some of the limitations that may hinder us from objectively analyzing the topic of the definition of life and, specifically, from investigating the issue of predictability and its dramatic consequences. I detail the mapping I made of some of my own personal human limitations and biases when I started to think of how to best address such a convoluted problem, and I further describe how I tried to go around these personal shortcomings of mine by utilizing a thought experiment, that is, by simulating in my head an experiment in my laboratory, with the usual careful analysis, but with instruments and tools we don't yet have. Specifically, our goal in this thought experiment is to dissect the brain and study its core software and hardware. I sketch what a thought experiment is and the many constraints on our thinking and objectivity, originating in our personal biases, our ego, and even our language, that such an experiment can help circumvent. Several bias-invoking theories such as the theory of the denial of death are considered. Timeless audacious thinkers such as Confucius, Einstein, Kant, Popper, Lao-tzu, Darwin, Copernicus, Freud, Schopenhauer, Wittgenstein, Kuhn, Magritte, Sartre, and Descartes, as well as numerous other artists, writers, scientists, and philosophers, are called upon to assist.

In the fourth part, I take a philosophical pause to emphasize what is really at stake when discussing human predictability. I begin by laying out the present limits of our scientific knowledge concerning death. Furthermore, the resemblance between humans and cyborgs, robots, androids, and humanoids is analyzed. I then present in simple terms the idea of a new definition for human (mental) life and death, through a measurable (quantifiable) observable, which is human predictability. Once reliable values of predictability are made available and we can measure predictability in a technologically mature manner, our understanding of life and who we are will be forever altered.

In the fifth part, I go back to the science of predictability. Even if one is unappreciative of philosophical interpretations, the science of predictability still stands. Here I continue the discussion from Part II and present several pathways that may counteract predictability, such as neural network noise. This is the beginning of a cookbook recipe for the pill. As noted, some advocate a reduction in decision-making variability, as it takes a toll in various human endeavors requiring accuracy and efficiency.[9] This is tantamount to advocating an increase in predictability. If brain noise can indeed counteract predictability, I argue in contrast to these advocates of less variability that a minimal amount of noise is a must if we are to maintain our humanity. In this context, I also discuss the quantum mind and the possibility of new laws of physics. Understanding the most fundamental physics standing at the base of all our brain building blocks is a must if we are to eventually see the big picture.

I then present a simplified model of how the brain interfaces between exploitation and exploration, a model that allows us to visualize the battle taking place inside the brain for and against predictability. I conclude by concisely stating the outcome of the thought experiment we performed.

In the sixth and final part of the book, I present contemplations concerning the consequences of the experiment and the new definition of human life for how we may view human behavior and how humans can use this newfound understanding to achieve transcendence. This transcendence is all about being able to produce the anti-zombie pill and being courageous enough to take it. I would not have written this essential part of the book if it were not for a clear directive by Eve.

Once I had escorted Eve back to her group and university hosts on that day that changed everything, and just before I said "be well" as we parted, she suddenly stopped at the door and turned around. She was now facing me. I can still see her perplexed face, and I can still hear her voice articulating the words quietly and slowly, as if she wanted to make sure I understood their importance. She said, "Once you understand the answer in scientific terms, please translate your answer into the words of the humanities." Now it was my turn to make a face, and I guess it showed bewilderment. She added, "It's not because I don't appreciate or don't understand the words of the scientific language. It's just because I don't think it holds all the necessary wisdom needed, if the human race is to successfully evolve and prosper."

In those first days after she left, the more I thought about it, the more it sounded like an oxymoron, a paradox beyond my skills, a task that would require of me, a physicist, to cut right through a Gordian knot.

Hence, while the previous parts were more tuned toward the accurate definitions and considerations of a hypothesis born of the school of the exact sciences, this final part of the book utilizes the language and contemplations of the humanities. Specifically, before sending her my final answer, I attempted to complete the construction of the bridge between the scientific and technological description of what is and that of the humanities of what ought to be. In fact, as I learned later, many people view this bridge to be an existential bridge for our civilization, as without this holistic approach, wisdom is lost. In any case, the more I understood the importance of her request, the more I strived to adhere to her clever directive.

I started by looking at what could be learned about human predictability from one of the most extreme human activities: war. I then found myself forced to also analyze the role of happiness, creativity, and even poetry, as well as art and architecture, and their implications for predictability and society.

Once my thoughts went as far as I believed they could go, I ended this part and the entire book with an epilogue, a personal final note to Eve. Here, I also make a connection between the new definition for human life and how philosophers have described what ought to be in terms of a transcending future human. Eventually, I found myself compelled to confess to Eve whether I feel there is any room for hope.

Confucius said that the person who asks a question may be a fool for a minute, but the person who does not ask may be a fool for life. This book, now in your hands, is a result of the mind-boggling questions that have bothered humans since the dawn of time: What is unique about us humans? What kind of freedom do we have? What may we aspire to become? Indeed, these questions have always been ferociously debated. However, the circumstances surrounding this conundrum, and indeed the entire backdrop for this crucial discussion, are now completely new. This book explains why we are currently living in dramatic times, as the understanding of quantum theory and in parallel the striking rise of the technology of AI and Big Data are about to allow us an extraordinary view into the very core of who we are and who we may become. As noted, without our being aware of it, the most profound experiment ever done on humans has already begun.*

Finally, a word of encouragement before we begin our journey down this rabbit hole: Although this book is aimed at a popular-level exposition of the topic, it includes quite a bit of technical detail, including a review of the most recent scientific work. To those who are somewhat disheartened by such detail, don't worry. It is not by chance that Einstein said that imagination is more important than knowledge. If you have a good imagination, that will suffice. In any case, the book is built in a modular way, whereby skipping dense technical parts is not a disaster, as simple concise overviews of the previous parts are intertwined throughout the text. Specifically, if you have no time or patience, then reading this introduction, as well as Parts I and VI, will tell you pretty much the gist of it.

We are now ready to journey into the battlefield in the depth of our brain, where the battle between random and deterministic, explore and exploit, is taking place, a battle that, according to our definition of life, is nothing short of being the most important battle we will ever fight.

---

*This book aims to portray in a concise and popular manner the main facts and arguments. For those who are interested in more detail, I present in comments and appendices a deeper account of some of the issues discussed in the book, such as empirical work on consciousness, the state of the art in human predictability, and so on. For those who require even more detail, the book includes hundreds of references.

Part I
THE PREDICTABILITY EXPERIMENT

1

The Near Future Will Change Everything

1. YOU CAN'T IGNORE THE RESULTS OF A WELL-PERFORMED EXPERIMENT

Dear Eve, I now begin my answer to your startling yet wonderful question. I have called it *The Human Test*. I hope you like the name. *The Human Test* tells the story of the most profound experiment ever conducted on humans, an experiment that has the power to provide us with clear-cut conclusions, an experiment that is already taking place, and astonishingly no one seems to know about it.

A predictive technology utilizing the all-powerful AI and Big Data, which, as I will explain, is on its way to know us better than we know ourselves, is being developed by our civilization for a variety of purposes (e.g., economic, security, and political), but now it is inadvertently about to teach us things about human nature that are far beyond what anyone could have imagined or intended. The earth-shattering results, which may completely change our understanding of who we are, are expected to be apparent and readily attainable already within our lifetime.

To tell the story of the most profound experiment ever conducted on humans, I would first like to explain what I mean by the word *experiment*, and specifically an *inadvertent experiment*.

To begin with, an experiment is any activity from which we may learn the fine details of some phenomenon, its causes, the laws by which it occurs and evolves, and its implications. The first striking thing to note about experiments is that you can't ignore the results of a well-performed experiment. You just can't. No matter how much you hate it, and no matter how much it goes against everything you believe, as long as it is a well-performed experiment, the results just can't be ignored. Darwin, for example, had such a hard time accepting the conclusions that emerged from the evidence, namely that the creation of humans was not an act of God (at least not in the way the Scriptures describe it), that he wrote to his friend that it was as difficult and excruciating as confessing to a murder.[1]

You can't ignore it because after several hundred years of doing experiments, we humans have learned that well-performed experiments speak the truth, or at least bring us closer to speaking the truth, or at the very least take us farther away from what is untrue. This is why being an experimentalist or, as some have called it in the past, an experimental philosopher is so fascinating, because you get to live next door to truth. Because of this I became an experimentalist. This and the love of physics.

Of course, at this point, one should have quite a few reservations against the use of the word *truth*. Let me note two.

The first has to do with the fact that what science believes to be true seems to have an expiration date. Some call it the half-life of facts, in reference to the term used to indicate the time it takes for one half of radioactive atoms in a sample to decay. For example, we once believed that Earth was the center of the universe. The 2013 book *The Half-Life of Facts* nicely describes the life cycle of facts. But science is well aware of this, and its stated goal of a never-ending journey toward the truth, layer by layer, lives well alongside the half-life predicament.* More so, science even insists on regularly attacking its own truths with new experiments as a way to continuously distance itself from what is untrue. In fact, science is so careful that it typically does not use the word *truth* but rather speaks of increasing confidence levels in, or corroboration of, some model representing our understanding of nature. This is why I noted above that well-performed experiments enable you to live next door to

---

*As scientific methodology improves, we are more and more aware of the reasons for the half-life of facts, and we are able to take more care. For example, the more complex a system is, the greater probability there is for hidden variables we are not aware of and the fact that correlation may not be proof of causation. That's why the half-life in physics is much longer than it is in biology, health, and medicine, for example. We are also increasingly aware of flawed statistical methods and ill treatment of error bars (uncertainties in data). However, while scientific methodology improves, we are always in danger of regression if scientists develop an agenda beyond the relentless pursuit of truth. Such agendas, for example, may be promoted by industry, which funds a significant part of academic research.

truth and not in the house of truth, as every good scientist knows that truth is infinitely complex and is never obtainable in its entirety. Consequently, a seasoned scientist is forever skeptical and understands that believing you have captured the entire truth is a fool's game. Anyway, next door is enough to blow your mind away with deep understanding.

The second reservation is frequently raised by philosophers and, frankly speaking, it is warranted. The philosopher Max Weber once said that science can explain what is but not what ought to be. Perhaps this is also what Bertrand Russell meant when he stated that everything has been figured out except how to live, and also what Ludwig Wittgenstein meant when he stated that even after we solve all the scientific questions, we shall still be without answers to the important questions. To put it another way, the biggest drawback of rationality, which is the foundation of scientific and technological thought, is that it cannot define the goals of humans but rather only the ways of achieving these goals. For example, we have seen how very rational people can become evil, or simply devoid of good values. If the word *truth* should include wisdom of life, or how to properly live life, then indeed science is missing out on the whole truth.* In fact, some would even argue that this is truly an existential gap that may eventually annihilate us. In this respect, one may refer to the *great filter* as an argument, where the great filter aims to solve the Fermi paradox of why it is so hard for us to detect advanced extraterrestrial life when there should be an abundance of it. Some versions of the great filter simply assume that technologically advanced civilizations self-destruct,† and

---

*It is perhaps worthwhile to briefly note that not only philosophers thought that this gap, between what science can deliver and what we actually need, is of crucial importance. Artists, spiritual leaders, and novelists all understood that it is key to our future and, as such, perhaps key to truth. For example, C. P. Snow's *The Two Cultures* speaks exactly of this gap, whereby the gap is between the culture of science and technology and the culture of the humanities; an extremely literate and clever person in one of the cultures is typically illiterate in the other. One culture is usually concerned with measuring and understanding what is, typically in terms of the physical laws of nature and how to exploit them. The other is similarly concerned with what is, whether through history, philosophy, literature, or poetry (as well as the study of economics and politics), but it is also concerned with what ought to be, typically in terms of the spiritual, ethical, and moral norms of humans and how they may be mapped onto human existence. The humanities approach the description of what is in a language that is also useful when trying to describe what ought to be, while the exact sciences do not. It is not by chance that Snow's book was listed as one of the one hundred books that most influenced Western public discourse since the Second World War; it seems people are indeed concerned that truth needs to be multidimensional and all encompassing, but not enough is done to bridge the gap.

†Regarding different explanations of why we don't see advanced extraterrestrial civilizations, see, for example, Robin Hanson et al., "If Loud Aliens Explain Human Earliness, Quiet Aliens Are Also Rare," *Astrophysical Journal* 922 (2021): 182.

if self-annihilation is proof of not possessing the whole truth, then science and technology indeed do not possess the whole truth.

The above reservations should be well received, but as long as one is aware of the boundaries within which the exact sciences operate, no one can dispute the fact that our knowledge is growing exponentially because we humans have finally managed to master a criterion for scientific truth, and this is a consequence of learning how to conduct and analyze experiments. The exponential growth of our fundamental knowledge is of course followed by a similar advance in our technology; while the former is typically hidden from the eyes of the public, the latter is evident and may serve as proof of the former.

It would be presumptuous of me to try to sum up in a few pages the glorious expanse of the philosophy of science, but I believe I have provided you with the gist of it: it is almost a miracle that we humans have managed to master experiments and the criterion for truth that comes with them.

There are different kinds of experiments from which we humans learn. For example, some experiments are naturally occurring. When an apple falls from a tree, this is not an experiment designed and built by humans, but still we humans can learn from it about gravity. There are, of course, experiments designed by humans for a specific task, such as the particle accelerators designed to find new particles, which will teach us why protons, neutrons, and electrons are as they are and how they contrive to construct atoms, the building blocks of nature and life. But sometimes we learn from inadvertent experiments. That is, we construct some technology or experiment for a specific goal, but we end up learning something completely different. A famous example is that of the microwave oven: an American engineer named Percy Spencer found out that a candy bar he had in his pocket melted when he was close to a microwave source. Another example is that of Bell Labs physicists Arno Penzias and Robert Wilson who built an antenna with the intention of finding radio signals from specific locations in the sky, but they discovered that even when they turned their antenna in random directions, they still found some annoying background noise, a kind of uniform signal. This became one of the most fundamental discoveries ever, because this signal, which came from everywhere in the universe, was found to be a temperature left from the Big Bang (this is now called cosmic background radiation).

The experiment described in this book is of the latter type, an inadvertent experiment. Eve, allow me to recap: The predictive technology utilizing the all-powerful AI and Big Data, which is on its way to knowing us better than we know ourselves, is being developed by our civilization for a variety of purposes. Now it is inadvertently about to teach us things about human nature that are far beyond what anyone could have imagined or intended. In fact, it could be that the results are already known to a privileged few, but we, the public, have not been informed.

2. MEASURING PREDICTABILITY

Dear Eve, before we go into the details of present-day technology and the inadvertent experiment that is already set in motion, let us indulge ourselves with a taste of what future instruments measuring humans may look like. For contrast, let us first begin with a highly unrealistic story (at least for the near future) and then move on to a realistic one.

Imagine you are walking with a great Nikon camera down Fifth Avenue in New York City. It is 9 a.m. There are lots of people walking in haste to work and lots of others, tourists, making their way slowly up and down the avenue. In your bag is a strange-looking lens. Your friend, an inventor, a secluded tech wiz, gave it to you last night and said that you should try it out on Fifth Avenue. He has recently become depressed but would not tell you why. Last night when you visited him, he could hardly speak. When you pressured him for the reason, he gave you this bizarre lens, clearly handmade, and told you it holds the reason for his demise.

On the corner of Forty-Second Street, you sit on a bench and put the bulky lens on your camera. Once your retina registers the first glimpse of what the camera is seeing, your hands automatically retract as if they have touched fire. You then lift the camera again and force your eyelids open. You feel a sharp pain in your temple, the same electric current surge you feel when you know you and your car are about to crash at high speed and there is nothing you can do about it. Your eye is telling your brain that you are seeing zombies. Walking zombies. Hundreds of them. Just like in those Hollywood films, only with less makeup and twitching. The lens somehow provides new information. More clarity. Their faces look almost human, but something, it's hard to put your finger on it, forces your mind to deduce without any ambiguity that you are watching zombies. Somehow, little things about their demeanor tell you that they are dead.

But different from the theme of those films, they are not out to get you. They even somehow seem harmless. They are just walking to work or behaving like tourists. Some of them are even laughing, some are hugging, and some are looking at their phones. But you have no doubt: they are not human.

When you take the camera down, your bare eyes again scan the avenue, and they tell your brain what they have always told it; all the people look normal again. After a few minutes you rush into the toilet of the public library. You are looking for a mirror. You then point your odd-looking lens at the mirror, and to your horror you find that you are also clearly a zombie. But that's impossible; you know you are normal. You recently had a thorough physical checkup, and everything was normal. Your blood, your heart, all normal. They even took your EEG and did an MRI scan because you said you had some recurring headaches. After a multitude of minutes in which you stare at yourself in the mirror, touching your face with one hand and holding the camera with the other, you eventually manage to move, and you make your way toward the house of your friend, the inventor of the lens. You want to give it back and never touch it again.

The above is obviously science fiction, not only because we don't have such a lens but because we don't even have a good definition of what a zombie is, so there is no way to interpret the results coming from any kind of physical measurement.

In contrast, let us imagine another scene that at first may also seem like science fiction but which, as I will explain, may be safely conjectured to be realistic in the not-so-distant future. Such a conjecture may be made based on an extrapolation of present-day technology and the identification of existing technological trends, as detailed below. But first, let us further indulge in the more realistic scene from the near future.

Imagine yourself in twenty or thirty years. Against their wish, the powers that be were forced to reveal that they have measured that most humans are predictable. You begin to wonder if you yourself are predictable and to what extent, so you decide to get tested. You make an appointment with one of those companies that for several decades now have been providing the service of scanning your DNA for health reasons or to find out about your family roots, only now you ask for the new service that they have started to provide only recently. You ask them to scan you for what has been commonly known as zombieness, or as the scientists simply call it, predictability.

In this future, the different opinions regarding the way in which zombies are dead have given way to a near consensus that it is probably connected to predictability. In the past, people would argue about what is dead in zombies. It is clear that their biology is functioning, so what about them is dead? Some people would say they have no soul, but *soul* is also not well defined. Others would say they cannot feel empathy and love, but they were not sure about this, as when zombies have children, they probably love them. So eventually most people agreed that the main point about zombies is that they are not free independent entities, and that's why in the movies they seem to move like a herd of sheep, or in more violent scenes, like a pack of wolves. Namely, they have no *free will* beyond automatic instincts and pack mentality. Consequently, in the future, most people view a zombie simply as a complex machine, a complex biological, chemical, physical machine, but a mere machine nonetheless; it is this that makes it dead, and this is why the commonplace jargon or street slang has equated the scientific term *predictability* with zombieness. But the connection to zombieness is of no real consequence. While in the near future people still find it hard to clearly define what a zombie is, or for that matter what consciousness or self-awareness are, they clearly understand what predictability is, and indeed it has been defined with mathematical precision.

"I made an appointment," you say as you approach the front desk. After a short wait in a luxurious-looking lobby, your name is called. "Are you sure you want to know?" asks the professional in a white lab coat. You nod. "Then please sign this waiver releasing us from all liability for the possible consequences of the knowledge which we will provide you." You sign the form. In fact, you have already ordered the examination and paid up front several months ago, and today you have only come to receive the result. They insist on giving you the results in person so that they may explain things to you and, more importantly, give you immediate assistance in case you need it. Quite a few people have tried to hurt themselves or simply fainted upon receiving the news, as no human wants to entertain the notion that they are a mere machine. Before you are handed a document with your level of predictability, and before you are allowed to leave, they give you a panic button. "We receive the signal 24/7 from anywhere on the globe," the man tells you. Although they are legally exempt, public pressure has forced them to take this initiative, as some people slowly drifted into deep depression. Some have simply disappeared, never to be heard from again.

Since you requested the test, their AI has been scanning the Big Data of the world for your digital footprint and has been analyzing your every move for the past decade. You have given them permission to access every piece of your private life, including all the data harvesting that the appliances and even the furniture in your home do, as well as your car, not to mention your phone and computers. Rumors say that their AI even finds ways to hack into street cameras and federal databases. Their AI has been building a model of you, and it has been comparing its predictions of your everyday decisions to the actual data. Namely, half the data was used to build a model of you by training the AI, and then the other half is compared against the AI's predictions of your decisions in order to quantify how predictable you are. It is strange, but these gazillions of bytes of information about you end up as just one number. P they call it, between zero and one, where $P = 1$ means completely predictable. A person with $P = 1$ is a machine, and if one agrees that a machine is a dead thing, then that person could consequently be considered mentally dead, though his or her body is alive. A simple single-digit number that has the power to change everything you ever thought you knew about yourself.

Jean-Paul Sartre stated that a person is nothing but the sum of his actions. The measurement of P follows exactly this philosophical wisdom, but in ways that I believe Sartre could not have dreamed of.

Mathematically speaking, the definition of P is quite simple. P describes how much data has to be collected on a human before their actions can be predicted. For example, if you had to watch your neighbor for years before you could make successful predictions about his or her preferences in future situations, then your neighbor should receive a lower value of P (i.e., he is less of a predictable machine). A higher value of P should be given if you could predict them after knowing him or her for just a few months.*

P = 0 means a person is completely random, as only complete randomness is truly unpredictable. I guess we would all agree that randomness cannot be the essence of life, as randomness means we are not able to truly think and decide on anything. For P = 0, any decision we make is just a roll of the dice. Thus, the definition of human life, or maximal human life, may be taken to be in between this randomness and the machine, namely, P ~ 0.5.†

Finally, one should note that the predictability these companies will be measuring in the near future is the predictability concerning goal-oriented situations in which we perform some task to achieve a practical goal. Most of our actions are of this type. However, one may create artificial tasks that are not of this type, such as asking a person to think of a random number. This is a task that has an inherently random nature where the only goal is to be as random as possible. As explained in Chapter 7, there are sources of randomness in our brain, and as explained in Chapter 8, in our two-layer model of the brain, some brains may be able to give dominance to randomness when required. Such randomness is not something that the AI serving these companies in the near future will be able to predict, but as noted, most if not all of our actions are not of this type. In the distant future, when technology could perhaps completely simulate the brain of an individual, then even the randomness of the brain may be predictable, but this is beyond the scope of the current discussion, which focuses on the near future.

---

*In more detail, P specifies the volume of information, V (where V is a positive number between zero and infinity, say in terabytes), we need to collect on a specific human being before we are able to predict his or her behavior with a reasonable level of success, say p = 90 percent (where lowercase p stands for the prediction success score). In more professional terms, P is thus simply the normalized p, namely p per unit volume V. The smaller the volume of information we need to collect for a good predictability score, say p = 90 percent, the more predictable the human is; that is, P is close to 1. Alternatively, for a fixed volume of information V, the prediction success score p concerning an individual will determine P.

†For those with some mathematical background, P, the normalized version of p, may be described by a simple qualitative model (just a guide for the eye), whereby for a high prediction success score of say p = 90 percent, P is described by $-\ln P = V$, or $\exp\{-V\} = P$. If V = 0, that is, we hardly need to collect any information about a human in order to predict him, then P = 1; namely, the person is as predictable as can be. He is a machine, and consequently he is dead. If V >> 1, i.e., V is very large, then P ~ 0. Namely, the person is completely random, and thus he is also not alive. To be alive, namely, P = 0.5, one would need V = 0.7. This is just an arbitrary number which in this simple model represents the optimal point.

Going back in time to our present day, we may ask: Could the instruments of the near future really measure predictability? My answer is that they surely will be able to do just that. The predictability of humans is a measurable observable that is already well on its way to being tested. As noted, at this stage it is being tested inadvertently. As described in the narrative above, in the future it may be tested intentionally, but at present it is still an inadvertent experiment. Nevertheless, it is about to uncover earthshaking new insights into who we are. Let us now go deeper into the topic of predictability.

Perhaps in the most pragmatic sense, the people who were most aware of the notion of human predictability in the past were the low-ranking officers of state security organizations who had to listen to hours of recordings that were secretly made of some individual, with microphones implanted everywhere he goes, like the officers of the East German Stasi. We can imagine that these eavesdroppers could very quickly sense how predictable a person is, and out of boredom they would eventually whisper his words before he even said them. But, very rarely, they would get a new subject, and after listening to him for a long time they would say to themselves that they have no clue what he is about to say next, even when the circumstances are quite ordinary. It is safe to assume that AI will soon completely replace those low-ranking officers of state security organizations.

In fact, as noted in the following, the AI constructions of today enable us to make quite a reasonable conjecture that companies and governments in the near future will know the exact thing a specific person will prefer when required to make a choice in a specific situation, whereby such a prediction will be made based on his recorded previous activities. Individuals will be predictable to a high degree, and the error in such predictions will become exceedingly small as the computing power and collected data become astronomical. The commercials advertising a product, and indeed the features of the product itself, whether materialistic, spiritual, or political, will be customized for each individual human being. We could say that AI will be the Stasi of the future in ways the Stasi of the past could not even dream of.

Let me briefly make a side note and mention that some experts are convinced that the next leap in AI's computing power will come from quantum computers. These computers, which currently have only a few hundred qubits (the quantum bit), are expected to rapidly become very strong, both in hardware and algorithms, as hundreds of companies are formed (some of them by my former students) and as billions of dollars are being invested, including by companies such as Google, IBM, Amazon, IonQ, D-Wave, Psi-Quantum, QuEra, Honeywell (Quantinuum), Pasqal, and many others. This is just one way in which quantum physics is expected to have a strong impact on the question of predictability. An even more profound impact is discussed in the dedicated section on the quantum mind in Chapter 7.

Let me also clarify that when I state that human predictability is a measurable observable, I mean by *measurable observable* exactly the same as when this term is used in experimental physics. Namely, that the parameter may be observed by a detector or an instrument and that the detector can assign a value (i.e., a number) to this parameter. For example, just like we measure speed, mass, or temperature, we will be able to assign a value, from zero to one, to predictability P. I noted above how this value will be measured; half of the data harvested regarding your personal choices, what has been termed your intelligent history, will be used to build a model of you by training the AI, and then the other half is compared against the AI's predictions of your decisions. This allows the calculation of P.

Acknowledging that predictability is a measurable observable is made easier when we realize that the science and technology of human predictability has grown significantly over the years, though it has still not attracted our attention (more details in Appendix C).

The predictability of humans and its potential uses were already recognized quite some time ago. In the early twentieth century, Edward Bernays took Freud's ideas (Freud was Edward's uncle) and applied them to the American market. He showed that it is very easy to predict and thus manipulate people. For example, smoking was taboo among women, and he almost completely turned it around. He began by convincing women that smoking equals thinness and that thinness is beauty. Appealing photographs with beautiful women smoking did the job. He then connected smoking with the emancipation of women by calling cigarettes "torches of freedom." All he needed to do was hire women to infiltrate a Fifth Avenue Easter Sunday parade in New York City, be visible so that the photographers could see them, and have a cigarette in their hand or mouth.[2] When the green color on Lucky

Strike packages did not appeal to women, he organized a social event at the Waldorf Astoria, a Green Ball, in which famous society women would attend wearing green dresses, and intellectuals were enlisted to give highbrow talks on the theme of green. Later on, he convinced the public that invading Guatemala was a good idea in order to fight against communism, while he received the actual request from a banana company that was losing money there.[3] It should not surprise anyone that the Nazi minister of propaganda, Joseph Goebbels, admired Bernays.[4] He was selected by *Life* magazine as one of the one hundred most influential Americans of the twentieth century.

More recently, numerous companies have appeared with the core business of making predictions concerning individuals. Such companies—for example, Behavio, which came out of the MIT Media Lab and was absorbed by Google in 2013—state that they deal with profiling and predicting human behavior. Behavio, for example, utilized data recorded by smartphones.[5] Another company, Affective Markets, claims to have cracked the human emotional DNA, enabling it to predict human preferences on a collective and on an individual level. They claim to have a system and method for prediction of musical preferences. In fact, still unpublished is their surprising claim for cross-discipline predictability, namely that through musical preferences they can score high in predictability on other human traits. They are now already claiming a predictability score of 80 percent for the choices of individuals concerning, for example, shopping. Predictability, this time regarding future divorces, is also being tested by people like John Gottman, an American psychologist and professor emeritus of psychology at the University of Washington, who has worked for over four decades on divorce prediction and marital stability. Such studies already claim a predictability rate of 80–95 percent.

Another example is Mind Strong, which claims to use powerful machine learning methods to show that specific digital features of individuals (i.e., our activities in the digital world) correlate with cognitive function, clinical symptoms, and measures of brain activity, allowing the company to predict mental performance. Ben Reis and his Harvard predictive medical group claim to be able to predict suicides years in advance (there are numerous such groups working on predicting mental health, e.g., Rosalind Picard at MIT). A new interdisciplinary field called digital phenotyping, utilizing available data from personal digital devices to profile and predict people, is expected to greatly empower the reliability of the predicting process. Data-harvesting sensors will be embedded in every walk of life (many of them under the innocent-sounding titles "smart—home, car, clothes"), and a growing number of companies will optimize the search for this data in the vast expanse of Big Data—Explorium, Babel Street, Rayzone, or Insanet being a good case in point.* The character of such companies will range from those which scan the information we choose to make public to those which will outright attack our privacy.† It has unfortunately already become a fact of life that organizations and countries are doing their best to collect as much data as possible in order to predict future behavior, not always with the most noble intentions in mind.[6] However, for the purpose of our discussion we are interested in capability rather than intent.

As noted, some have started to call our digital footprint "intelligent histories." Perhaps the most in-depth data harvesting will eventually be done by the online gaming industry (for example, currently more than a billion people are already playing online). To be more and more attractive, they will need to study each and every one of us in great depth. Social media (such as Facebook [now Meta], Instagram, Twitter [now X], TikTok, Netflix, Spotify, YouTube, and even news organizations) are also becoming extremely good at harvesting data and predicting what content you will engage with. For the most part, algorithms for deciding which posts to show us (which they call recommender systems) have been optimized for that.[7] These algorithms can see exactly what content you spend time on or react to. Even you, the user, cannot describe your behavior as well as they can. This data is fed into deep-learning algorithms that use the huge amount of data with billions of data points to make predictions and output the next content. It is important to note that these algorithms are learning machines intended to learn more and more about you, not only through regular data harvesting but also through analyzing and grading how good their past predictions leading to recommendations were. The experts call this self-evaluation the objective function. I have not found reliable numbers stating how good these companies are in making predictions. Perhaps they themselves are afraid to publish such results.

---

*See also projects like KnowItAll by Oren Etzioni at the University of Washington.

†See also *The Age of Surveillance Capitalism* by Shoshana Zuboff of Harvard Business School. She also speaks of behavioral-prediction products. Speaking of unbalanced capitalism that may be amoral or perhaps even immoral, it is interesting to note that some scholars conjecture that Adam Smith, considered by some the father of capitalism, and who at the same time expressed much interest in morals (he considered his other work, *The Theory of Moral Sentiments*, to be superior), would have disliked what has become of his free market idea, as in terms of morals it leaves much to be desired. I believe data harvesting for economic gain will push this lack of harmony between the two philosophies to the extreme.

In the near future, data harvesting is going to skyrocket also because of two additional trends, wearable sensors and personal assistants. Wearable sensors already include smart glasses, smart watches, and even smart rings (e.g., the Oura and Galaxy rings). Of course, the demand for privacy will require that the data be stored safely, but digital data is never completely safe against malicious hackers, or agnostic leakage. AI assistants will start by performing relatively technical tasks such as attending meetings for you, as promised in the immediate future by, for example, Otter.ai. However, our wish to improve our well-being or simply our efficiency and success in life will mean that a growing number of us will extensively use personal AI assistants. But for them to do their job, they must have an endless stream of data points about you. They will need to know you down to the tiniest detail. Many people are already embracing this future. For example, at the turn of the millennium, Gordon Bell started digitizing a significant portion of his personal life, and later, together with his colleagues from Microsoft Research, he wrote the book *Total Recall*. Life logging has actually already turned into a community called the self-quantification movement. To be successful, personal AI assistants will even start harvesting data about you in an active manner by life logging, or journaling, your life, not only through passive sensors but also by prompting you with queries about your experiences and wishes. For example, this is where a chatbot named Pi, from a 2022 company called Inflection.ai, is going.

Let us briefly note that farther into the future, data harvesting will also be done directly from the brain itself, from outside the skull and eventually from within. From the outside, systems like EEG and MEG will continue to be developed by academic researchers and by commercial companies such as Neurosity and gTech medical engineering. Even Meta has announced that it is working on interpreting MEG data. Brain implants are also gaining momentum, with companies such as Neuralink, Synchron, Blackrock Neurotech, and Inbrain Neuroelectronics, or for academic EEG measurements, the so-called intracranial EEG (iEEG). Research into this technology had already begun in the 1970s under the heading of *brain-computer interface* (BCI). The concept of cyborgs (humans with enhanced capabilities due to advanced implanted hardware) will, in the more distant future, push brain implants to the extreme. However, I expect that long before such implants become mature data-harvesting devices, the external devices noted above will become so omnipotent that they will be good enough for high-level predictions of human actions.

At present, so much data is already harvested that the industry has started to seriously contemplate how to manage and use it. For example, one idea is to put all the data collected by the different companies and apps in one well-suited repository. Of course the claim is that this very secure place will be owned by the user, and it will only be used for the user's benefit.[8] Obviously, good-willed, conscientious AI developers speak of "AI alignment" with human values, which they also call "value alignment," in order to ensure that AI and Big Data are always used for the benefit of humans. Will history judge this heroic effort as being too little, too late? However, we are not here to discuss the dangers and ethics of social media and other platforms but rather how far their algorithms have gone and how far they will go in the near future in proving our conjecture of predictability.

Present-day academic articles and theses reflect the interest of universities and research institutions in these questions. Recent titles include "Predicting Human Preferences from User-Generated Content," "A Machine-Learning Approach for Predicting Human Preference . . . ," "Predicting Human Preferences Using . . . ," "Choosing Prediction over Explanation in Psychology: Lessons from Machine Learning," and "A Big Data Revisit to Fundamentals of Personality Psychology."* These are real titles of existing academic papers and theses. As part of this trend, a 2010 Northeastern University popular statement declared: "Human behavior is 93 percent predictable, a group of leading Northeastern University network scientists recently found."† Such studies will clearly become more and more prevalent as industry and the state understand the potential and start investing heavily in related research.

To the aid of these deep-learning algorithms, which analyze the unique data produced by each individual and can therefore come to know and predict that specific individual, will come more and more advanced universal models of human behavior.[9] As these models become increasingly more detailed and accurate in classifying people (e.g., through categories of personality types and traits), the algorithms will be able to use knowledge about

---

*One of the concluding statements of this PhD thesis by Gal Ben Yosef (2019) is that "the predictive approach to personality modeling could theoretically lead to models that render human behavior extremely predictable."

†Obviously there is a huge gap between the popular declarations and the detailed scientific articles, in this case *Science* 327, no. 5968 (2010): 1018–21, in which one finds that this work focused on mobility, and that for individuals they could only reach 80 percent predictive accuracy; https://cos.northeastern.edu/news/human-behavior-is-93-predictable-research-shows.

choices made by other individuals in the collective ("tribe") we belong to in order to predict us.* These universal models of human behavior will enable the deep-learning machines to reach a higher level of predictability of an individual while requiring a smaller volume of information that needs to be collected regarding the actions of that individual.

Will we get to 100 percent predictability ($P = 1$) on all human activity and decision-making fronts? Answering this question is an arduous task. As far as we know, the capability to predict us is only now making its first baby steps. However, taking into account all that we already know, the evolutionary trajectory seems quite clear, and it is now a matter of making sound extrapolations. Perhaps the hardest challenge for AI when it comes to learning from past decisions of an individual and predicting any future choice will be to understand the context surrounding a human when they make a decision.† The emerging research and technology field of context awareness is addressing exactly this challenge. Most importantly, the exponential growth of three capabilities suggests that the problem is not insurmountable and that within a relatively short time AI will also understand the human context. These are data harvesting (including wearable sensors), detailed human models (either through theories of personality types or through collaborative filtering), and enormous computing and memory power. The motivation to solve the context problem is already extremely high and will grow tremendously in coming years. It will be required not just for the successful implementation of the personal assistants mentioned earlier but also for AI to be able to play its part in our intimate life. Chatbot companions geared for intimate conversations will become more and more sophisticated, and they will require understanding of context in order to be effective. The 2023 outcry of users over the change of policies introduced by Luka Inc. to their Replika app teaches us just how advanced these intimate companions have already become. The outcry made it clear that true bonds between humans and AI have formed. Similarly, the emerging field of erobotics, in which AI is geared for developing emotional and sexual bonds with humans, will clearly push the understanding of context to currently unimaginable capabilities.[10] These

---

*For example, with the help of something called collaborative filtering, in which the key idea is to use the data of people similar to the specific individual in order to predict what he or she will do (e.g., Chapter 3 in *Recommender Systems Handbook*).

†The relevant context is mainly made of the physical, relationships, and intent contexts. For example, the exact weather conditions at the time of a decision making are part of the physical context. See Appendix C for more details.

trends allow us to make an educated estimate that in the near future the context will become clear to AI.

Finally, I should note that the idea that AI will be able to predict our most intricate thoughts, such as goals, emotions, and desires, has already over the years been given numerous names by the different professional communities. For example, technical literature refers to the ability to predict one's beliefs concerning a particular situation as the theory of mind. Similarly, in the 1990s, Rosalind Picard, already noted above, talked about affective computing, which would be able to predict one's emotions.

To summarize, the preliminary data coming in regarding specific forms of predictability already observed allows us to conjecture with a relatively significant degree of confidence that a high level of predictability will be found in humans. It also seems likely that in some people, predictability will be measured to be close to 1, while in others less so, so that predictability will be found to be a personality trait (already now, some scientists notice a correlation between the level of predictability and certain personality characteristics[11]). We discuss in several chapters and sections of this book what processes may limit predictability, for example, noise and randomness. These will form the base for the anti-predictability (or anti-zombie) pill, which is discussed in detail in Parts V and VI.

In any case, it is quite clear that the experiment to test our predictability has been put into motion, and not a thing can stop this gigantic ball from rolling down the hill. It is nothing less than astonishing that the most profound experiment ever conducted on humans has begun mostly under our radar.*

However, for the deep-learning machines of AI to be able to fully predict us humans, it is not enough that they have as input a significant amount of our "intelligent history," namely, data harvesting that was done on our actions in our life so far, and that they know quite a lot about general human behavior, and it is not enough that they possess unimaginable computing power. For the ultimate predictability of a human to be possible, for P to actually be an as-

---

*As noted in the introduction, I sincerely hope that when the time comes at which a high level of predictability is found, we will indeed be told of this milestone, as it constitutes a crucial event in the evolution of humans. Unfortunately, as described, I fear that different interests may create strong opposition to making this public.

tounding 1, our mind must be a complete machine, that is, a fully deterministic entity, where determinism simply means that there will always be the same output for the same input. There are no buts or ifs in the matter. There must not be any "free will." This is crucial. If the atom is a machine and our brain is nothing but atoms, it stands to reason that the brain is a deterministic machine as well, and then AI can achieve the ultimate predictability over humans. To clarify, AI does not need to know what's inside our skull, namely, how the brain is built. It does not need to simulate the neurons and the neural networks that connect them. As far as the AI doing the predicting is concerned, the brain is just a black box with inputs and outputs (namely there is no knowledge of what's inside). For any specific brain, AI is simply given the task of predicting future decisions taken by a brain based on that brain's past choices. The point made here is that for AI to fully predict the output of this black box—based on its previous outputs—the black box must be a machine. If there is some true randomness in the black box—for example, some natural random number generator (e.g., some kind of biological roulette), or what mathematicians would call a choice sequence—then even the strongest AI will not be able to achieve the ultimate predictability score of 1 for any individual. This crucial question is explored in the next parts of the book.

3. THE HUMAN MACHINE

Eve, as the implications are so grave, you must be wondering if it's really true that the brain is deterministic. Indeed, the implications are earth shattering; once we accept that predictability is machine and machine is death, we can start talking of machine death. If their predictability score, P, is close to 1, people may just as well be declared machine-dead, even if their heart and mind are functioning. In this section I track the origins of human predictability and begin to estimate what normalized predictability score P most humans would receive from those instruments of the near future.

But before going forward, I remind you that our goal is to answer the question, What is the definition of human life? Similarly, we may ask, In what sense am I alive? or, What is the meaning of my life? Or, Am I really alive? Here we set about distilling this intuition that haunts us into a rigorous reflection.

We are by far not the first to think about the machine that is within us. Many have had the suspicion that we are machines. Perhaps even Freud, who said that we are controlled rather than in control. I recall, for example, W. G. Sebald's book *Austerlitz*, wherein Marie wants to wish Austerlitz congratulations and happiness from the bottom of her heart, but she knows that by doing so she is really just wishing for the machine that its mechanism will work smoothly. Putting order in these thoughts and scrutinizing them in a methodological manner is now our goal.

Let us begin by looking at our brain. Much is already known about the brain. New machines like fMRI and EEG are making available an abundance of data and insight. However, it is quite clear that we are still in a state where what we don't know is much greater than what we do know. We do know that the brain is made of protons and neutrons (about 10^{27} of them), which form the atomic nuclei, as well as of electrons that surround each nucleus and enable chemistry to unfold and make molecules, which in turn enable the construction of almost one hundred billion neurons and 10^{15} connections in each human brain. These neurons have all kinds of parts, such as the dendrites and axons that enable communication through the small gaps between the neurons that are called synapses. We also know a lot about which regions of the brain are responsible for which activities. Our rate of accumulating new knowledge is growing so quickly that it would not be far-reaching to assume that in the next one hundred years we will learn more than we did in the past ten thousand years. We know so much about the brain, and yet we know so little.[12] (For those who like quantum language, we may say that we are currently in a superposition of knowing a lot and at the same time knowing so little.)

To think of the brain while we are still in a state of significant ignorance will require us to really stretch our imagination. Richard Feynman said that we must stretch our imagination to the utmost, not, as in fiction, to imagine things that are not really there, but just to comprehend those things which are there, as reality is so unbelievably complex.*

Let us therefore be bold and, in view of our reflections concerning mental death, ask the most important question of all: Is the human mind deterministic? Namely, can its present state be predicted by some outside agent based

---

*An interesting anecdote is that Nobel laureate Feynman was not admitted to Columbia University because of their quota on the number of Jews admitted. It is perhaps also interesting to note that Feynman, who is credited with discovering why the *Challenger* exploded, was rejected by the army and eventually not recruited on the grounds that he failed his psychological examination (some claim he faked his mental instability). One may wonder if Feynman's "lack of normality" is not tantamount to "lack of predictability," which enabled him to make this and other discoveries. See also sections on creativity and divergent thought.

on its state in the past, even just in principle? Although we have already introduced the term *deterministic*, I emphasize again that it means that specific initial conditions will always lead to the same outcome sometime later. Situation (state) A always leads to situation (state) B. Identifying specific initial conditions thus allows one to exactly predict the future of that specific system. For example, if we let a ball that is initially at rest roll down a thirty-degree slope on a planet with a well-known gravity, we know exactly at what velocity it will be after, say, ten seconds. It will always be the same answer, giving the same final velocity.

So, is the human mind deterministic? If it is, it is predictable, at least in principle. To dive into the depth of this question, we must completely isolate the brain from input. We should also make sure there is no light or vibration or any other form of noisy disturbance directly perturbing the brain itself. Only those 10^{27} protons and neutrons, and a similar number of electrons that make up the brain, are now of interest. And all these particles are moving about according to the influence of forces from the other particles, namely their immediate environment (the distance between protons and neutrons in the nucleus is typically one femtometer, namely, one millionth of one billionth of a meter, and the distance between atoms is typically one Angstrom, or one tenth of a billionth of a meter).

Hence, if we put into some imaginary supercomputer the position of each particle in the brain at a certain moment (say we have some super-MRI able to image the position of each particle, thus giving us the exact initial conditions of the brain), and the known interactions between the particles, the computer will be able to analyze the position of each particle one hour later or a day or a year later. This implies that if we believe that our thoughts are nothing but an outcome of a specific arrangement of these particles simply because there is nothing else there, then our thoughts may be anticipated by the computer in advance. For example, if the particles are arranged one way, then I like blue, and if in another way, then I like red, and the same goes for any thought or decision we have or make. It's all about the specific arrangements of the particles in the brain. I must emphasize that when I say that "our thoughts are nothing but an outcome of a specific arrangement of these particles" and that "there is nothing else there," I mean it literally. That is, all the hereditary influences from our DNA and genes, and all the influences of our upbringing, including all traumas and fears, and happy memories and education, and psychological needs and wants, are all already represented in the specific arrangement of the particles, as all these past influences have created the specific arrangement that exists now.

Of course, in a real-life situation, our brain is also influenced in real time by the environment outside the brain, like input from the eyes and ears, but in principle we can also take into account the position of all particles around us, and so the outside influences can also be calculated and determined in advance. If we believe Einstein, that things cannot influence us faster than the speed of light, then only those things that can reach us at the speed of light during the time of the experiment need to be taken into account. Physicists call this the light cone.

Our imaginary supercomputer will have to take into account only four interactions of each particle with its environment, as there are only four forces in nature: the electromagnetic force, the strong force that keeps the nucleus together, the weak force that gives rise to radioactive radiation from the nucleus, and the gravitational force.* We know how to exactly calculate the dynamics of the particles under the influence of these forces, so in fact a system made of such particles and interactions is one big machine, similar in principle to a mechanical clock.

Obviously, even 10^{27} nucleons and a similar number of electrons, each with its degrees of freedom like position and momentum and spin (not to mention the internal structure made of quarks), multiplied by the interactions between each particle and its immediate neighbors, is a daunting number hard to fathom, which is in fact larger than the number of stars in the known universe, and may inhibit, even in the far future, such a calculation. But as we note in the following, it could very well be that our supercomputer will need to take into account only neurons or even clusters of neurons to be able to predict the brain. That's already a much smaller number.

Before continuing with our imaginary experiment, I briefly remind you that, as noted at the end of the previous section, in order to expect that ulti-

---

*There is of course an intensive search ongoing continuously for new additional forces, but so far nothing definite has been found. Such searches are, for example, taking place at CERN, where the Large Hadron Collider, a giant particle accelerator deep underground, is supplying new fascinating data on the interaction between particles. An interesting anecdote is that some people say that CERN is the closest humans have come to God. When I was doing night shifts there as part of my PhD, I sometimes had the same feeling, not only because our knowledge of the universe went so deep in that place, but perhaps mainly because in that place so many people from so many different countries, races, and religions found a way to work together in harmony.

mate predictability of humans by AI will be achieved in the near future, we only need to know that in principle the black box that we call a brain is deterministic. We don't really need a supercomputer or AI to follow all the particles inside the brain, or even clusters of particles, a feat that may be achieved in the distant future. For our extrapolation of what the near future will look like, our imaginary experiment and the idea of the supercomputer that we have just described are only required in order to come to the conclusion that in principle the black box in our head is deterministic. That would be enough to theoretically allow AI to predict us simply based on our past actions, without knowing anything about what is inside our skull. To put it bluntly, for all AI cares, when it comes to analyzing our actions, this black box that we call the brain might just as well be filled with mashed potatoes.

Going back to our imaginary experiment and the question of determinism, if there is nothing but atoms in our brain, the brain seems to really be made of some mechanistic clockwork, indicating that there is no difference between the single atom with which scientists work in the lab, and which we typically deem to be a machine, and the mountain of atoms that we call the brain. It may very well be that the difference is only in complexity but not in essence. Of course, most of us would agree that we have consciousness and the atom doesn't,* but this seems not to prove anything concerning determinism, especially when *consciousness* is not well defined.

But what about free will? Doesn't free will contradict determinism? I will devote a section to free will, but for now let me just state that it could be that the complexity of the machine is merely imitating free will, as currently it is too hard for us to practically predict the machine's outcomes. Namely, the mere fact that our peers constantly await our decision on different issues, without any possibility of predicting what our decision is going to be, gives us the feeling that we are free agents, even if we are not. That's what I mean when I say that complexity is imitating free will.

---

*Some try to claim that atoms are conscious as well, but as someone who works with single atoms on a daily basis, I dare say that such a claim requires emptying the word *consciousness* of the meaning we typically ascribe to it. See, for example, Jonathan Moens, "Are Atoms, Bacteria and Plants Conscious?," *ScienceLine*, 2020.

I will, in the following, also devote a section to the soul, a popular argument among many to try to defeat the idea of the machine that is us, as the soul, believed to be a free agent, is said to ultimately control the machine.

Some people may even argue that our irrationality proves we are not deterministic. Indeed, rational behavior is more predictable. For example, it is easier to predict a rational (logical) player in a chess game. Although random behavior is not rational behavior, irrational behavior may not be random. It may very well be that a deterministic process gives rise to irrational behavior. For example, in the book *Predictably Irrational* by Dan Ariely, a behavioral economist, he asserts that we are systematic and predictable, though irrational to a large extent. The origin of some of this irrationality is in the difficulty in rationally analyzing the tremendous amount of information we receive, so evolution developed mental shortcuts. But these shortcuts have nothing to do with indeterminism (the latter means nondeterministic, the opposite of determinism).

Though, as noted, I discuss free will and the soul in the coming sections, I would like to state here that to the best of my knowledge there is currently no clear-cut proof that they do not exist. However, obviously, proof of predictability, as may become available in the very near future, will put to rest all these vague notions of having some free part within us, where by "free" I simply mean nondeterministic, namely, unpredictable by some external agent. It goes without saying that a high level of predictability is also incompatible with many other intuitions we have about our unpredictability, like involuntary outbursts of emotion or uncontrolled, subconsciously driven decisions. So, again, measuring human predictability has the power to resolve many open questions we have about ourselves.

Let us now go back to our brain. Recalling the description of the current technological state of the art in the previous section, we must try and understand why already now, with our rudimentary AI technology, human predictability seems to be quite high. Could it really be that our brain is deterministic? Pierre-Simon Laplace, the great eighteenth-century French thinker who worked on everything from math and physics to astronomy and philosophy, already made this same argument when he spoke about the whole universe. He said that the only reason for its state at this very moment is its state in the past, and the only reason for its state in the future is its state in the present. Namely, everything is completely deterministic. Some people call this supercomputer I was talking about the Laplace demon.

If the Laplace demon does exist, we may at this point be contemplating the following conundrum: If our brain is indeed deterministic, can a deterministic system, namely, our brain, prove that it is deterministic? My answer would be that it can, if it has been predetermined to do so by the laws of nature according to which the physical state of the universe is evolving.

And yes, one may contemplate in bewilderment the fact that if determinism is complete, then my meeting with Eve as well as the writing of this book have been predetermined ever since the Big Bang, which is when the ball of the natural universe, with all its elementary particles and atoms, started to roll.

Some people are even more extreme in how they view this determinism. For example, once Einstein proved that time is a relative thing, people started thinking that the past and future are not objective things, and they began talking about the possibility that the past and the future exist now, but we just don't know how to access them. If that's true, one should in principle be able to play some tricks—like some fast spaceship that changes your frame of reference or going through wormholes in space-time—in order to actually observe in the present also the past and the future. They call it the block universe, in which everything, including the future, exists now in some corner or dimension of the universe. If this is so, then of course everything is predetermined, as the future has already happened.

However, Eve, let us put this extreme notion of the block universe aside and focus on standard determinism in which there is a clear arrow of time. This is the scenario of the Laplace demon. I must tell you that I have always felt that this demon sounds like some grand conspiracy against humans, a conspiracy against humans not contrived by other humans but by the cosmos and its laws of physics. I obviously use the word *conspiracy* with a negative connotation, because if the Laplace demon is real, it is a huge blow to our potential as free beings. In simple terms, I hate this demon.

But even the Laplace demon may have its limits. For example, what about uncertainties, like the quantum uncertainty principle, which states that we cannot know everything about a system with infinite accuracy? Doesn't this mean that there is no determinism, as the initial conditions cannot be known accurately?

One can address this argument in several ways. First, some scientists claim that these uncertainties required by quantum theory only reflect the state of our knowledge of nature and not the state of nature itself. Second, quantum theory is based on randomness, and if brain functions really work according to quantum principles, then our thoughts are based on randomness, and we have already noted that complete randomness is the other extreme, just as bad for humans, at least humans as we would like to see them. So assuming that the brain is dominated by quantum rules is not going to save us, at least not on the face of it.

But most importantly, there have been quite a few calculations done showing that on the large scale of many neurons, the brain does not work on quantum principles. I dedicate a section to quantum theory and the quantum mind later in the book, but here let me just briefly state that quantum theory only works well on the micro-scale of very small, well-isolated systems, and scientists[13] have calculated that the number of atoms involved in the firing of even a single neuron, and the strong interaction with the liquid environment of the brain is enough for quantum theory not to be relevant.*[14] According to today's standard theories, quantum states simply cannot survive in such a multiparticle ambient environment, composed of liquids at room temperature.

Let me clarify that when I say that quantum theory only works well on the micro-scale of very small and well-isolated systems, I mean that otherwise it suffers from a destructive process that kills the quantum nature of the system. It can either come from the coupling of the environment to our small system (which is akin to the environment continuously measuring our small system), and that's why we need the system to be well isolated, or

---

*Let me already whet your appetite by saying that quantum mechanics may still have some surprises up its sleeve (see Section 7.2 on quantum theory). For example, some people such as Nobel laureate Roger Penrose believe that quantum processes have a real influence on our decision-making process, and they use terms such as *quantum consciousness* or the *quantum mind*. Scientifically speaking, there is no proof that the quantum mind exists, but as we continue to study quantum mechanics, there may of course be new surprises. As described in this book, some hints are starting to come in. In principle, such far-reaching hypotheses must be commended, as they are what drives science forward. See Stuart Hameroff and Roger Penrose, "Consciousness in the Universe: A Review of the 'Orch OR' Theory," *Physics of Life Reviews* 11 (2014): 39–78. See also Roger Penrose, *Shadows of the Mind: A Search for the Missing Science of Consciousness* (1994). Finally, in Sec. 7.2 I bring several references claiming that the calculations done showing the brain cannot be quantum had errors, as well as references suggesting quantum processes that are more robust.

it can originate from within the system if the system is large and has many particles in it. This is in many ways like asking whether the brain is similar to the spaceships that take us to the moon or to Mars, or the *Voyager* spacecraft which went the farthest. These marvels of technology must suppress quantum uncertainties and randomness in order to fulfill their designed mission. They must be completely deterministic. But they are built of quantum atoms, so how do they do it? Namely, how do they suppress the quantum nature of the spaceship? It is not from the coupling to the environment, as the environment is empty space (the few photons or hydrogen atoms that hit the ship can make sure its position is well defined but not its internal workings). So what makes sure its internal workings are deterministic? In simple terms, we may say that it is through the fact that there are so many atoms and thus so many parallel quantum processes within them, each for another atom, that they simply average out to give no uncertainties and no randomness.* This average of many random processes is so accurate that we may call it deterministic, and that's why our marvels of technology work so well. In the same way, current calculations show that even the firing of a single neuron requires the involvement of too many particles, and the quantumness is averaged out very accurately. Hence our previous statement that quantum theory should not be relevant in the brain.

So, if quantum theory is indeed irrelevant, we are left with the classical theories in which there are no inherent uncertainties, and we can know everything as accurately as we want. The infamous Laplace demon reigns once again. The bottom line is that, in what concerns the brain, it currently seems that the prevailing theory is the old classical deterministic theory with the four forces we mentioned. There are many more arguments to be made in this debate on a deterministic brain. For example, some claim that the field of thermodynamics, with its concept of entropy, puts a limit on how much information may be available in the universe, so that not everything can be known, but this may not be relevant to our discussion on this tiny piece of the universe that we call a brain. Still, as already noted, 10^{27} particles is indeed a daunting number, and one would need a huge amount of information to know the initial condition or to follow the evolution of the particles involved.

---

*The suppression of the quantum nature of a system is referred to by physicists as decoherence, and the suppression of uncertainties as collapse. We deal with both in the section on quantum theory. But for now, the term *averaging* is good enough.

Here two comments may be made: First, as noted, it could very well be that the supercomputer will need to take into account only neurons, or even clusters of neurons, to be able to predict the brain, so the required amount of information would be much smaller. Second, these are "technical" considerations regarding the possibility that in the far future a supercomputer can actually do the calculation. But, as explained, our goal here is only to decide if the brain is deterministic in principle and not if our technology can ever actually follow its internal clockwork.

One may also bring up the theory of chaos, where minute changes in initial conditions can bring about staggering changes (divergences) in the final state of a system. A simple example is the three-body problem[15] (see also chaotic pendulums), for which there is no general closed-form solution, meaning there is no general solution that can be expressed in terms of a finite number of standard mathematical operations. The motion of three bodies is generally nonrepeating, except in special cases. The Laplace demon or the supercomputer will never be able to follow that, as heavy numerical calculations are required (as opposed to the simpler analytical calculations), and in any case, even if the latter is not a problem for our hypothesized supercomputer, the exact initial conditions could never be known with the required accuracy to defeat chaotic divergences, simply because of the limitations of our measurement devices. However, it stands to reason that the theory of chaos cannot play a major role in the brain. The brain has to function under continuous fluctuations of temperature and temporary imbalances in the concentrations of sugar, salt, and other chemicals. If the theory of chaos was dominant in the evolution of the state of the brain, our thoughts would really become random. There must be some mechanism that stabilizes the system, like an error correction code in a quantum computer, based on redundancy. Otherwise, we would not have been able to exhibit the rather impressive stability of our thoughts.

Some may argue that human psychology is a metaphysical process (i.e., beyond the laws of physics), and consequently the supercomputer will never be able to predict the brain's future state based on its current state, as there are no mathematical equations that may describe it. They will then go on to argue that as psychology is an output of the brain, if you can't predict psychology and show it to be deterministic, it cannot be that the brain is deterministic. In light of this argument, an interesting question is whether this metaphysical psychological state creates changes in chemicals like hormones, or is it the other way around, namely that the chemicals arriving at the neuron clusters are controlling our psychology. If it's all about chemistry, the work of the supercomputer is much better defined, as we understand what a chemical state is much better than we understand what a psychological state is. For chemicals, we have equations just as we have for atoms. One may find a sea of scientific literature indicating what the answer is. These articles state something like this: The ever-growing recognition of the importance of hormones and chemicals that serve as neurotransmitters in determining human behavior is directly influencing new schools of thought in psychiatry, neurobiology, sociology, psychology, and political psychology. The ability to closely track changes in the state of neurotransmitters in real time is allowing scientists to decipher mental states and diseases, turning them into physiological states and diseases. One should not rule out the possibility that feelings such as hate, love, and friendship are also a mere outcome of neurobiology (namely, the chemical and biological state of the brain), with all that this implies.

The chemicals of love, such as dopamine and phenethylamine (PEA), are examples of how our psychological state is determined by our chemical state. The increasing recognition of the dominant role of these chemicals, and the growing knowledge of the clockwork (mechanistic) functionality of the brain, is what is driving psychology departments to move from the faculties of the humanities to the faculties of exact science. Highly-read journals with names like *Molecular Psychiatry* have recently started appearing.* Cause and effect in the brain, the hallmark of a deterministic system, can be measured!

The fact that even consciousness itself has a chemical basis is evident from the well-known observation that we can shut consciousness down with chemicals. It's called general anesthesia, and it is practiced in every hospital. Quantum biologist Luca Turin noted in 2014 that the only thing we are sure about regarding consciousness is that it is soluble in chloroform. We may assume that this is a first step in learning about the clockwork machinery behind consciousness. By trying different anesthetics, and even different isotopes (same atom or molecule but with different mass, as neutrons are added), we should expect to learn about what kind of processes in the brain (e.g., electron transport or spin physics) stand at the base of consciousness. For example, it

---

*The journal states that it "publishes work aimed at elucidating biological mechanisms underlying psychiatric disorders." In 2021 it had the relatively high impact factor of 16.

is claimed that different isotopes of lithium give rise to different mothering behavior in rats, and different isotopes of xenon (an anesthetic) produce different levels of unconsciousness.[16]

—⚏—

Finally, even if we are not satisfied with our present-day knowledge of what is really happening in the brain at the level of atoms or chemical molecules, or whether the hypothesized supercomputer will be able to follow it, determinism seems to show up on many other scales as well. As we leave the tiny scales and look at the brain as a biological system made of giant clusters of neurons, there is more and more evidence for determinism, or no free will, at this mega-biological level. If there is determinism on the scale of clusters of neurons, then the work required by the supercomputer in order to predict our thoughts will be much simpler.* At this higher level, the geometric structure or connectivity maps of a specific brain determine its output.[17]

Some theoretical frameworks are already being put in place to describe the machinery of the brain in a deterministic manner on the large scales of brain regions and the interactions between them. An example is the recently developed theory of the global neural workspace described in Appendix B. Obviously, as noted in the introduction, every theory that has the eventual goal of describing with an ever growing accuracy the action of the brain by some mathematical function, such as IIT mentioned earlier or the free energy principle of Karl Friston,[18] supports the notion of a deterministic brain, as mathematics is deterministic! (as long as there is no stochastic [random] term built into the mathematical equation).

—⚏—

We may now take a long look at our brain. It seems it is a machine, as we found that there are several layers enforcing determinism. Namely, even if by some miracle we rid ourselves of the never-ending external brainwashing by the political, religious, and business sectors (which of course increases the

---

*If working with large clusters of neurons is enough to follow the clockwork of the brain, this negates the thermodynamic argument noted above, as well as the problem of quantum uncertainties, which would not exist for large structures such as a neuron or a cluster of neurons. The requirement for a super-MRI able to image the position of each particle, thus giving us the exact initial conditions of the brain, would also be considerably relaxed.

predictability of our brain), the internal determinism still works on several levels. There is the Laplace demon working at the level of the elementary particles, running the whole clockwork on the four forces of nature. But even if this demon is weak and faulty, for example, because of quantum uncertainties, there still seems to be no freedom of choice, as the brain is controlled by its chemicals. There also may be no freedom of choice on much larger scales, as the structure of the individual brain is determined by its genetic makeup, and modern theories, which are based more and more on mathematical formulas, equate this structure to that of a machine, which is in principle deterministic. We further discuss this lack of freedom in the following, for example, in the section on free will.

Before we conclude this first view of the brain as a machine, it should be clarified that suggesting that the brain is a deterministic machine is by no means equivalent to suggesting that all brains are the same. Every brain machine is obviously quite different. Indeed, recent research has identified that the temporal behavior of each brain is so unique that it has been equated to a fingerprint.[19] But this uniqueness of structure and function says nothing about determinism. It could just mean that each of us is a different machine; each of us is in a different jail.

To conclude, there is mounting evidence supporting a variety of arguments explaining why the human mind is deterministic. Therefore, one may deduce that as the power of AI increases, and based on our "intelligent history," the outcomes of the human mind will become predictable in the near future, even if the inner mechanisms of the black box that we call the brain are not fully understood for a very long time to come.

More specifically, the complexity of the brain, and consequently the arguments for and against the determinism of the brain, is such that it will most probably be impossible to fully resolve this question of determinism and complete predictability by using the well-known arguments, some of which have been presented here. It also stands to reason that more and more accurate probing into the exact clockwork mechanisms of the brain, as will certainly become available, will not, in the foreseeable future, answer the question of determinism. Furthermore, even if the brain is deterministic at large scales, it could be that we still won't have enough computing power or memory to do a full calculation, that is, to create an accurate simulation of the decision-making process, enabling us to predict the outcome, namely, the choice we will make.

In this debate over the determinism of the brain, there is currently no decisive argument and no smoking-gun experimental result, and none is expected in the near future. The game-changing potency of the new predictability score, P, is that it is a straightforward, unambiguous observable that, through a feasible measurement—that is, of the outcomes of the brain—will be able to put this debate to rest, as a measurement of a high predictability score P would prove that the brain is deterministic.

Can we conjecture that many if not most humans will be found to have a very high value of P, that is, close to 1? If the brain is deterministic, then the answer is yes. As described in the previous section, it stands to reason that in the near future each of us will be able to be tested for our value of P.

As the topic of whether the brain is deterministic is of crucial importance, in the following chapters we continue to analyze this question, as well as to explore the mind-boggling possibility that the value of P may vary for different individuals, each having a different brain.* Finally, we will also examine the prospect that if an individual receives a high P score, he or she may yet fight to change their fate of doom.

If you are out of time, you may choose to skip to the bottom line presented in Part VI.

---

*One may already be asking oneself about this scenario of varying levels of predictability in different brains: What would happen when a less predictable person meets and interacts with a predictable person? Does the former not reduce the predictability of the latter? I would divide external influences into sustainable ones that have a great impact on our brain (even to the point of brainwashing) and random stochastic influences such as noise in the street or on the bus, or birds and planes flying overhead, or random fluctuations of temperature and light intensity, which seem to have little effect on our core behavior (see discussion above on chaos). The brain seems to be very stable against such "noise," and the reason is probably straightforward: to create real mental changes in the brain, you probably need real physiological changes to take place, and these take time to form, and they eventually form only once a consistent long-term effect is applied to the brain.

But what if an unpredictable person has a long-term interaction with a more predictable person? I suspect that if he or she is truly unpredictable, they will have a negligible effect on such physiological changes in the predictable brain, as they will be pulling this brain at random in different directions, and this simply emulates noise.

However, as explained in the following parts of the book, there may indeed be ways to influence a predictable individual toward less predictability.

Part II

IS DETERMINISM IN THE BRAIN REALLY POSSIBLE?

2

Common Arguments against the Possibility of Predictability

1. EMERGENCE VERSUS REDUCTIONISM

Dear Eve, let us start our in-depth look under the hood of the human machine, as it were, by reviewing some of the common arguments against the possibility of determinism, which as noted is required if complete human predictability is to be achieved by AI and Big Data. A first interesting argument against the determinism of the human mind, or at least against a determinism that leads to predictability, may be found in the philosophy of emergence.[1]

The point of view that everything we see could be explained by just computing the dynamics of all the elementary particles at play, most importantly atoms, is called reductionism. Perhaps the simplest example is that of a guitar string. In every undergraduate classical wave course, you learn that by taking an infinite chain of small masses (atoms), you can arrive at a wave equation that describes the features of the entire string, from which you can know the sounds (frequencies) it will make. The periodic table of elements itself may be seen as a manifestation of reductionism, as from the properties of single atoms we can infer the properties of the bulk material, for example, whether it will be electrically insulating or conductive, whether it will conduct heat, and what its optical properties will be. Indeed, the importance of atoms as the fundamental mechanism that drives everything was emphasized by many great thinkers. Feynman, for example, said, "If, in some cataclysm, all of scientific knowledge were to be destroyed, and only one sentence passed on to the next generation of creatures, what statement would contain the most information in the fewest words? I believe it is the atomic hypothesis (or the atomic fact, or whatever you wish to call it) that all things are made of atoms—little particles that move around in perpetual motion, attracting each other when they are a little distance apart, but repelling upon being squeezed into one another. In that one sentence, you will see, there is an enormous amount of information about the world, if just a little imagination and thinking are applied."

But is reductionism always true? The philosophy of emergence claims it is not. For example, from the knowledge we have of water molecules, can we predict the shape of snowflakes or fractal water crystals forming on glass, and can we, by knowing the atomic composition of sand, predict the exact ripples seen in sandy surfaces hit by winds or currents? Some claim we can't, but as far as I know there is no proof of that. It could be that again we simply lack sufficient control to follow all the degrees of freedom (e.g., position and momentum of all involved particles), as the amount of computing and memory needed are huge. It could also be that we don't understand accurately enough the interactions between the particles. It could even be that for some of these complex systems such as snowflakes, we would need quantum computers or simulators, as the dominant effects have a quantum nature.

The history of emergence is very long. Aristotle, for example, felt that "the whole is other than the sum of the parts."

It is very hard to prove either way. As a counterexample to reductionism, in 2009 a far-reaching paper tried to prove the existence of a system you cannot understand with reductionism.[2] They specifically claim to prove "that many macroscopic observable properties of a simple class of physical systems (the infinite periodic Ising lattice) cannot in general be derived from a microscopic description. This provides evidence that emergent behavior occurs in such systems and indicates that even if a 'theory of everything' governing all microscopic interactions were discovered, the understanding of macroscopic order is likely to require additional insights."

In his 1972 paper "More Is Different," Philip Anderson[3] claimed that multicomponent physical systems can exhibit macroscopic behavior that cannot be understood from the laws that govern their microscopic parts. Some philosophers, such as Michael Strevens, agree. He states that "the high-level sciences neglect low-level mechanisms for principled reasons, and will continue to do so even in their finished form. They need not, and indeed should not, draw on the lower-level sciences for their explanatory content, nor need they be constrained by the lower-level sciences' explanatory organization of things."[4] In our language, we may say that by lower-level mechanism or sciences he simply means interactions at the microscopic level of a few atoms and molecules, and by high-level mechanisms or sciences he means the description of the behavior of very complex systems including a humongous number of particles of different kinds.

Many scientists, however, including prominent figures such as Stephen Hawking, do not agree. Hawking noted that as soon as all fundamental laws of the universe are understood, we will in principle be able to explain all macroscopic phenomena. I should again emphasize that we indeed need understanding but also enough memory and computing power to follow all the different particles in a system as they evolve through time.

The above thoughts, that physical systems can exhibit macroscopic behavior that cannot be understood from the laws that govern their microscopic parts, come under the point of view coined as emergence. As philosophers Brigitte Falkenburg and Margaret Morrison put it, "a phenomenon is emergent if it cannot be reduced to, explained or predicted from its constituent parts. . . . Emergent phenomena arise out of lower-level entities, but they cannot be reduced to, explained nor predicted from their micro-level base."

Emergence is the opposite of reductionism and, in the mind of some people, may indicate that you cannot fully calculate complex systems, therefore making them unpredictable in principle. Consequently, the argument may be made that a supercomputer calculating the dynamics of all elementary particles in a brain may still not be able to predict the future state of that brain.

Again, as far as I know, there is no proof that reductionism is not correct.* However, in my view, even if reductionism is correct (which is my personal

---

*It is also interesting to note that belief in reductionism is not the same as believing we know all the required physical laws of nature. In fact, if one accepts the notion of an infinitely complex reality, it is clear that new laws of nature are yet to be discovered, and we will need to use them when we come to calculate a complex system such as the brain. The fact that new physics is around the corner should be evident not only from the above philosophical argument regarding infinity, but also from open questions that the scientific community is currently working on, such as the nature of dark matter and of the dark energy that accelerates the expansion of the universe. New physics should also appear once we are able to access more dimensions (e.g., at least ten dimensions are hypothesized by string theory). Finally, it should be noted that in his book *Shadows of the Mind*, Roger Penrose suggests that only new physics can explain consciousness. See Section 7.3 on new physics.

guess), emergence is still very important, as it points out that because of our finite memory and computing power, we should at different scales be using different theories for the best scientific efficiency, namely, minimal resources for the best predictive power. Specifically, the theory of emergence teaches us that at each scale we need to use laws that can give us relevant results with fewer degrees of freedom, as in statistical mechanics where an insightful example is the gas equation $Pr = nRT/V$. We don't need to follow every particle in the gas in order to know its important properties, such as pressure (Pr) for a certain amount of material (n = number of moles), at a certain temperature (T) and volume (V), where R is a constant.

Indeed, it is this scientific efficiency heralded by the philosophy of emergence that I believe points to the fact that even in a deterministic reductionist world, different scales may require the use of different laws requiring less degrees of freedom, and it is this that will eventually, in the far future, allow us to follow the internal clockwork mechanisms of the brain, even with the finite memory and computing power available in that future. So, while some would claim that emergence works against the predictability of the brain, I would claim the exact opposite. It heralds the possibility that we will find equations that can predict the brain at the macroscopic level, namely the level giving rise to brain outputs, that is, decisions. Indeed, as described in different parts of this book, this is exactly where the field of neuroscience is heading.

Be that as it may, we are again confronted with a highly complex debate within philosophy and science. But the beauty and strength of our measurable observable, predictability P, is that even if this debate over emergence, as indeed expected, continues to linger even in the far future, the possible earth-shaking near-future implications of P hold firm.

Indeed, if in the near future a high level of human predictability is found by AI and Big Data, this will contradict all previous assumptions regarding the existence of some source of free thought, lack of determinism, or unpredictability. Specifically, measuring a high level of predictability P for humans will eliminate the possibility that emergence is a source of some freedom or indeterminism.

2. FREE WILL

We continue with our efforts to estimate what predictability score P most humans would receive from those instruments of the near future, AI and Big Data, as described in Part I. The existence of *free will* is clearly a force against predictability. It is a breaker of the machine, but do we possess it?

Our day-to-day experience convinces us we have free will. After all, we make decisions all the time. For example, an architect once told me in defiance, "You know, every time I design a living room and I decide whether to make it round or square, my free will makes the decision."

Free will was promised to us by many, starting with religious scholars, even in the face of determinism brought about by a God knowing all in advance. For example, Maimonides, the Jewish philosopher of the twelfth century, said, "What are the commandments for if everything is predetermined? There is so much punishment for sins; what is that about if we don't have a choice?" He then explained that only nature, which is external to man, is predetermined, like the stars and the rain. He is not alone in making such a distinction. Specifically, a school of thought popular among some philosophers, called compatibilism, states that free will is compatible with determinism.*

But how can we assume something other than the idea that it's the same nature inside our skull as outside? How can the latter be predetermined and the former not? What physical evidence or rational intuition hints in this direction? There is none. If you consequently believe that the same nature that is outside the skull is also inside the skull, then the only way that free will is compatible with determinism is if you engage in some fancy philosophical acrobatics, which, for an empirical physicist like me, unfortunately implies that you dilute the operational meaning of these words to the point where they don't mean much.

A similar tension arose in the Eastern religions such as Buddhism, Hinduism, and Taoism. This time, the deterministic forces of cause and effect are called karma, which in Western terms might be called destiny. How can karma be compatible with free will? Just like Maimonides, religious scholars of the East had to find a way out, and of course they did.

In the 2010 edition of the *Stanford Encyclopedia of Philosophy*, under the title of "Moral Responsibility," it is asked, "Can a person be morally respon-

---

*Among the proponents are Thomas Hobbes, David Hume, and Harry Frankfurt, and, more recently, Daniel Dennett, Frithjof Bergmann, and John Martin Fischer.

sible for their behavior if that behavior can be explained solely by reference to physical states of the universe and the laws governing changes in those physical states, or solely by reference to the existence of a sovereign God who guides the world along a divinely ordained path?" Obviously, if you want a world in which moral responsibility reigns, you must allow for free will, but is this proof of the existence of free will? If anything, it is just an additional source of bias when we come to discuss the issue.* Scholars like Spinoza, considered to be one of the great rationalists of the seventeenth century, and who trained himself his entire life to overcome the biases of the mind, voted against the existence of free will.†

Some present-day neuroscientists, such as Kevin Mitchell, in his book *Free Agents*, join those who promise us free will, while others, such as Robert M. Sapolsky, in his book *Determined: A Science of Life without Free Will*, are firmly against.

So what might the answer be? One thing seems to be clear: We live in dramatic times in which we will soon have a quantitative answer, as predictability cannot be compatible with free will, and the measurement of predictability is just around the corner.

But before we continue on the crucial topic of free will as an argument against determinism, allow me to demonstrate and emphasize how powerful

---

*Similarly, Nobel laureate Anton Zeilinger, the president of the Austrian Academy of Science, with whom I did my postdoctoral fellowship in Innsbruck, stated, and rightfully so, that if everything is predetermined and there is no free will for the scientist, there is also no point in doing science, as our predetermined questions may very well give rise to a false picture of nature. So the motivation of all forms and schools of thought to believe in free will is huge, and consequently the potential for bias when we come to discuss the question of free will is enormous.

†Physicists of course also have their opinions. In 1980, a theoretician from CERN, John Bell, who worked extensively on some of the strange features of quantum mechanics, such as "spooky action at a distance" (i.e., two entangled particles affecting each other at superluminal speeds, namely, speeds greater than the speed of light, something that is not allowed by the theory of relativity), stated that if we indeed have a deterministic world, namely, no free will, and if we add to that some predetermined correlations between the state of the mind of the researcher and the state of the measured system, a situation which he called superdeterministic, we can even get rid of the weird stuff in quantum mechanics. Also, the measurement problem, concerning how quantum systems register as normal (classical) information in our measuring devices and our brain, goes away. There is no randomness left in the theory. Perhaps all this hints toward determinism and no free will? Let me also mention here that one of the fathers of quantum theory, Erwin Schrödinger, had the intuition that the randomness and uncertainty involved in quantum theory do not suggest free will or even indeterminism. See E. Schrödinger, *Nature*, July 4, 1936, 13.

determinism in nature is, how A always leads to B. We are so used to cause and effect that we never really stop to wonder about the overwhelming consequences. So let us once again inhale into our intellectual lungs the Laplace demon and then continue with the consequences.

Nature follows very strict rules: for example, conservation of energy, which tells us how high we can throw a stone; conservation of momentum, which governs the scattering of billiard balls on a pool table; and conservation of angular momentum, which gives rise to the stability of the Frisbee and the bicycle. And, of course, everything also follows Newton's laws of motion, such as the one stating that any force acting on an object, divided by the mass of that object, equals its acceleration. But let us exemplify the precise laws that run the world with the simplest law of them all, Newton's first law, the law of inertia.

(Eve, I allow myself a little bit of physics here, but if you have no patience for that, please simply hop to the next paragraph.) When you sit in a car and the driver turns the car to the left, your body tends toward the right-side door. That's because your body wants to persist in its original direction, as required by Newton's law of inertia, while the right-side door wants to go left together with the rest of the car. It's also why air around an atmospheric pressure-low goes counterclockwise in the Northern Hemisphere of Earth.* Similarly, if a rocket is fired directly north toward a target that is directly north of us, it will hit east of the target. This is called the Coriolis force. It is also referred to as a fictitious force, as it is really nothing but a consequence of the law of inertia. Astronauts float in spaceships that are in orbit around Earth not because there is no gravity but because of the centrifugal force, which is another fictitious force due to the law of inertia. As noted, the law of inertia is the simplest of the deterministic laws governing our universe, and it has endless mesmerizing impacts on the world around us. Another example is the Foucault pendulum. It is proof that Earth is turning. Because of the law of inertia, these very tall pendulums always want to point in the same direction of space, but since the fact that Earth is turning doesn't

---

*If something flows north, it comes from a region of high side-velocity toward the east, because of Earth's rotation, to a region of low side-velocity, because while the rotation is still once in twenty-four hours, the radius in the north is smaller, and hence the velocity eastward is smaller. So air coming from the south would bend to the right, namely to the east, because its velocity toward the east is faster than that of Earth, and air coming from the north would also bend right for the same reason, namely westward.

bother them, they do something very weird. If we stand on the floor under these pendulums, we turn together with the floor, which is turning together with Earth, but as the pendulum keeps the same direction in space, it seems to us as if the pendulum is continuously shifting its direction of motion. The first such pendulum was positioned in Paris in 1851. Typical pendulums have a rope length of about twenty to thirty meters, with a ball at the bottom weighing about one hundred kilograms.

The bottom line of these forces of nature is that everything is so sharp, so predetermined. Cause and effect work perfectly, and for every set of initial conditions, we know what the final outcome will be.

In a simplistic manner, we could say that what we are actually doing when we take the precise physical laws of nature of the outside world and apply them to our brain in charge of our mental capacities is to describe a law of mental inertia. This is of course not a real law of the exact sciences with some mathematical equation, but rather a philosophical proclamation that represents the idea that the nature that exists outside our brain and the nature that exists inside of it are one and the same.* Hence, the use in the title of our law of the word "mental," which comes from the realm of the complex outcomes of the brain, and the word "inertia," which comes from the realm of the exact sciences, seems fitting. Consequently, by *mental inertia* we simply mean that our thoughts are a product of the deterministic laws of physics, and there is nothing outside this mechanistic point of view.

Speaking of mental laws, let us contemplate for a moment the idea that, regarding what happens inside our skull, one may invent all kinds of "laws" to describe the human experience. As noted, these are obviously not real physical laws of nature, which come with mathematical equations, but rather insights into human behavior hinting at the core code of our operating system.

---

*In fact, the idea of taking laws of nature from the realm of physics in order to understand or describe other realms is not new. In 1636, the English philosopher Thomas Hobbes met the astronomer Galileo Galilei, famous for his description of motion, and started to outline how society could be described by laws of motion and the mathematical tools of physics. A similar influence can be found in the works of eighteenth-century philosopher David Hume, who was influenced by Newton and talked of "mental forces" and the "force of habit," which is reminiscent of our law of mental inertia. Hume even talked of the force of habit as being analogous to gravity.

COMMON ARGUMENTS AGAINST THE POSSIBILITY OF PREDICTABILITY

One may, for example, take the law of mental inertia further and state that, just as with Foucault pendulums, it represents the difficulty the human brain has when it comes to changing habits or opinions. Like the pendulums, we continue to point in the same direction even if the ground is turning. In this case, if we go back to those instruments of the near future that will measure the value of P for each individual, the more mental inertia we have, the less information they would need to collect about us in order to predict our future behavior in a variety of circumstances. Thus, the more mental inertia we have, the closer P is to unity, and the more (mentally) dead we are.

We can also invent the law of chains, which states that any mental inertia you find in another human or even another animal is also to be found in you, unless you have managed to prove otherwise. I think it was Rosa Luxemburg who said in the beginning of the twentieth century that those who do not try to move do not feel the chains.* Indeed, there are strong chains in each of us giving rise to mental inertia. The mere fact that most religious people are devoted to the same religion their parents were is but one example of many regarding these mental laws in the sphere of human activities.

Or we can invent the law of increased faith, which says that if your belief that something is true grows stronger with the passing years, namely, your mental inertia grows stronger, make sure it is not because of your fear of discovering that you have been living a lie, and in many ways squandered your life away.

These mental laws are simply a fun way to represent more scientific theories of human behavior or personality models, which, as we noted, may help AI and Big Data score a higher P value for humans, but these laws or personality models are still far from proving determinism or the lack of free will.

Attempts to prove the existence of a mechanistic, predictable, deterministic human mind, though very rudimentary, were in fact already made a long time ago. Perhaps the most famous of these early attempts was that of Sigmund Freud.

He propounded rules about how we are built and what really drives us. For example, we may mention that boys yearn to kill their father and sleep with

---

*I guess we should not be surprised to hear that eventually she was jailed, tortured, and murdered (perhaps we could find some hope for the future in the fact that there is a street with her name in Berlin).

their mother, and girls have penis envy. These are obviously highly simplified examples from his very complex theory.

Freud of course failed to prove the accuracy of his predictions. In fact, he was smart enough to keep his predictions quite vague so that they could not be clearly contradicted. Generally speaking, everything a man desires consciously may be attributed to a subconscious Oedipus complex, and it may be shown that they are not incompatible. But it is exactly this vagueness that puts Freudian theory outside the realm of science. It is not a science but a pseudoscientific theory, which may perhaps be called a strengthened dogmatism, as it is constructed in such a way that makes it impossible to contradict it, namely, to refute it by some empirical test. It is interesting to note that similar considerations have led people such as Karl Popper to the idea of refuting Marxism as a science.

So Freud and Marxism do not prove determinism, although they aspired to. But even if individuals and the collective are still not fully scientifically predictable through Freud or Marx, it may simply be that the number of relevant variables is still too large for science and technology to comprehend them all, to measure and integrate computationally this wide space of parameters, without which we cannot build a scientific or technological procedure that has predictive power over humans. AI and Big Data are going to dramatically change that.

And perhaps it's not only the large number of parameters that did not enable Marx or Freud to come up with exact models. Perhaps we still don't understand which parameters are really relevant. This is also the suspicion in fundamental physics. Giants like de Broglie or Einstein thought that the theories are not complete, and some aspects are still hidden from us and from our measuring devices. Some scientists even use the name *hidden variables*. But as we noted, as computing power and the available data become humongous, predictability is becoming better and better in all disciplines. In fact, social predictability is becoming so high that some even call it social physics (see Section 9.5 on society).

Some people may claim that the mere fact that we have survived is proof of free will, as each of us, from the dawn of time, has had to make many independent decisions to ensure our survival. For how can a predetermined mind react appropriately to so many varied challenges? But deterministic is not predetermined in the simple sense that there is a library of situations telling us how to react to each one. Deterministic only means that everything follows very precise rules. Namely, one can view our survival in a completely different light. Evolution has created a feedback loop designed to survive, like an autonomous car will learn to survive on the road. In this sense, is our brain different from that of a lizard? Evolution has built a machine that works according to the laws of nature but at the same time is capable of appropriately reacting to challenges from the environment.*

What is a feedback loop? Nowadays we use it in technology all the time, for example, for temperature or laser frequency stabilization. At the core of a feedback loop are three elements: first, a sensor that constantly monitors the parameter you are interested in; second, the electronics and computer that identify deviations from the preferred value of this parameter and decide on a correcting signal; and third, the actuator, which can apply a physical action to the system according to the correction signal, thereby bringing the system back to its desired state. In fact, each of us has primitive feedback loops at home that go into action when we set the temperature in our oven or refrigerator, or the air-conditioning system.

Human feedback loops are sometimes studied under a different, less intimidating name. They are called decision-making processes, an interesting field of cognitive science.[5] Other names, more intimidating, at least in my view, are the stimulus-response model, or the stimulus-organism-response (SOR) theory, where the latter has been widely used in studies of online customer behavior.

Feedback loops can also learn. We see it in our labs. We call it machine learning. Feedback loops are of course deterministic. They are just an algorithm. The brain may be viewed as a powerful feedback loop. For example, near the end of the nineteenth century, an American psychologist, Edward Thorndike, put cats in a trap, placed some fish outside the trap, and showed us how the cats learn to escape. Every time he would put them back in the trap, it would take them less time than previously to escape. The mind repeats

---

*It also stands to reason that to react effectively to challenges from the environment, evolution also emphasized and enhanced the logical mind (see Section 9.4 on logic). Some cognitive scientists, like Anil Seth, would tell you that we are so fine-tuned to survive that "we perceive the world not as it is, but as it is useful for us"; https://www.theguardian.com/books/2021/aug/25/being-you-by-professor-anil-seth-review-the-exhilarating-new-science-of-consciousness.

what works for it. I believe he called it the law of effect. Today some may call it reinforcement learning. Responses that produce a satisfying effect in a particular situation become more likely to occur again in that situation. This is exactly what learning algorithms do. In this sense, Darwinism can simply be looked at as the survival of the fittest feedback loop.

The above learning by trial and error may become very advanced. For example, studies of decision-making processes talk about exploit versus explore. Typically, when we know how to get a reward, we act on it—namely, we exploit the situation—but sometimes scientists find that even if the environment (stimuli) is extremely clear and simple, decision-making processes make the wrong decision. They call this lapses in judgment. Some speculate that this is not a bug in the system but rather a smart way to further explore the situation in order to better understand it, and perhaps later on get even bigger rewards. In the science of artificial computer neural networks, this is called the search for the global optimum (minimum) (akin to a global minimum or maximum of some function, called the extremum, i.e., in terms of the human, the best solution), as sometimes the system is steered by the feedback loop toward a local optimum (i.e., namely, a good solution but not the best possible one). See the section on noise in Chapter 7 for further insight on this.

We may say that the human feedback loop runs on a chemical algorithm in which humans are driven, just like any other animal, to some chemical balance in their brain (or chemical equilibrium).[6] We can safely assume that robots will have such (not necessarily biologically driven) self-learning feedback loops in order to survive.*† So, survival is not proof of free will.

Of course, the human feedback loop even goes beyond learning and survival. It seems to show intent beyond mere instincts of reaction to the challenges of the environment. But as we note above and in Section 9.2 on happiness, this intent comes from a feedback loop responding to internal brain states that are out of chemical balance, with the goal of restoring balance. The state of the brain in which chemical balance is restored is dependent on the genetic makeup of the brain and changes that the environment imprinted in the brain ever since it was born (e.g., education). This creates long-term intent in the feedback loop.

---

*To get a feel for just how advanced feedback loops can be, one may read the story of DeepMind's AlphaGo Zero, a deep neural network that was able to win Go (Chinese chess) matches against the world champion. The neural network learned the game by playing against itself.

†More on robots in Chapter 6. There, among other more serious things, in a footnote, you can also find a bizarre personal anecdote concerning the root of the word robot.

More recently, different approaches to the question of free will have been taken. For example, two mathematicians from Princeton, Simon Kochen and John Conway, came up with a free will theorem which states that if scientists have some free will in the sense that their thoughts are not determined by the past (namely, by some Laplace demon), the elementary particles should have the same free will or free behavior. This in fact strongly hints toward an answer to the question of what the difference is between a single atom and the cluster of atoms that we call a brain. (See Section 7.2 on the quantum mind.)

People have also tried experimenting with free will, for example, the seminal experiments of Benjamin Libet in the 1980s. Libet demonstrated with EEG measurements that conscious intentions are preceded (by some five hundred milliseconds) by a specific pattern of brain activation, typically called readiness potential (RP), suggesting that unconscious processes, not some conscious free will, determine our decisions, and we are only retrospectively informed about these decisions. However, it is also commonly accepted that he showed that even if we don't have free will, our consciousness at the very least has a veto right. I believe he called it "free won't." This seems to be halfway to free will. But this is controversial, and many believe that free won't is also a prerogative of the subconscious rather than the conscious mind. What is not controversial is that he proved, by measuring brain activity before the subject declared his conscious intention to perform an action, that decisions originate from the subconscious, not from consciousness. Namely, we don't control our initial decisions. Many experiments have followed the Libet experiments, and in recent years the debate over the veto right and its meaning continues.* In principle, Libet's results were confirmed by several other techniques as well, such as fMRI and intracranial recordings. For example, in 2013, neuroscientist John-Dylan Haynes of the Bernstein Center for Computational Neuroscience Berlin and colleagues had volunteers decide whether to add or subtract two numbers while in the fMRI scanner. They found patterns of neural activity that were predictive of whether subjects would choose to add or subtract that occurred seconds before those subjects were aware of making the choice. Neuroscientist Sam Harris has concluded from such findings that we are "biochemical puppets."

---

*The two-layer model of the brain presented in Chapter 8 nicely illustrates the veto right.

On the other hand, in 2012, in a book titled *The Will and Its Brain: An Appraisal of Reasoned Free Will*, Hans Helmut Kornhuber and Lüder Deecke—who discovered in 1964–1965 the RP, or as they called it, the *Bereitschaftspotential* (BP), the brain potential in the EEG that, as noted, precedes all our willed movements and actions—state that a person has "relative freedom," which may be increased or decreased by self-improvement or self-mismanagement, respectively. We will revisit this crucial hypothesis in the following. Recent publications emphasize that the jury is still out.[7]

As fMRI, EEG, and similar probes become better and better, I expect we will have more and more data, but the complexities of interpreting the data will most probably remain. A rather comprehensive list of experiments that followed the Libet experiments may be found in the Wikipedia entry "Neuroscience of Free Will," where it is stated that "the field remains highly controversial." Let me also briefly note that the engineers of recent models of brain functions and consciousness have also tried to touch upon this explosive and very complex issue. For example, the proponents of integrated information theory (IIT, see Appendix B) have tried to argue for free will by saying that "we—not our neurons and atoms—are the true cause of our willed actions," but of course "we" is ill defined, and even these authors found it necessary to end their work by emphasizing a disclaimer, saying that the "argument for free will hinges on the proper understanding of consciousness as true existence."[8] But in my view, this is just an unfounded attempt to prioritize a cluster of atoms (the brain and its output, which we humans named *consciousness*) over single atoms, endowing it as the "true" reality, and in this sense it follows the line of thought of the philosophy of emergence. In fact, as I already noted concerning the free energy principle and other attempts to describe the brain with mathematical precision, I believe that all such efforts, which will clearly continue to intensify, must eventually succumb to the idea of a deterministic brain, as mathematics is deterministic! (as long as there is no built-in stochastic term).[9]

Finally, I briefly add that some people confuse creativity and free will, and they view our human creativity as proof of free will. This is not so. In Parts V and VI, I return to the important question of creativity.

Let us end this brief exposition on the excruciatingly complex topic of free will with a bottom line by saying that even if one is not able to scientifically resolve one way or another the argument over free will and determinism with existing methods of probing the internal clockwork of the brain, especially since there is no quantifiable definition of what free will is, the brain outcome we observe in the form of human actions tells us a significant part of the story. Namely, the criterion for whether freedom exists must be in the size of the space of actual human choices, of real actions, that is, in the variance (diversity) of realized human behavior. (By the way, companies are already measuring this diversity. For example, Spotify measures listening diversity and calls it the generalist-specialist [GS] score). Of course, one may want to hypothesize that we humans choose, out of free will, not to undertake diverse actions, but this is an ad hoc assumption that seems unwarranted.*

In other words, when examining the question of freedom, we cannot allow ourselves to be blinded and deceived by what seems to be the potential of freedom, namely, by what we think is theoretically possible, or what we believe is allowed by external constraints such as the laws and the norms of society. All this is too complicated and too vague. The only thing that matters is the variance of realized human behavior, namely, what we actually did at the end of the day. And even though it seems that this variance of actual human behavior is quite small, it is still hard to draw conclusions at the level of the individual brain. But in contrast, if a high level of predictability P of the individual brain will indeed materialize in near-future measurements, then this whole discussion over free will will have come to a dramatic end.

In such a case, it will become evident that free will does not exist, as predictability and free will are incompatible. It will then become clear that complexity is imitating free will simply because it is hard for our peers or

---

*To explain what I mean by an ad hoc assumption, let me turn to philosophy of science. There is a general consensus among scientists that if certain predictions made according to some general theory have been refuted, then if the theory can only be saved by adding ad hoc hypotheses that have no independent supporting base of their own, the general theory should be discarded. For example, the old Ptolemaic theory according to which the planets move in circles around Earth did not compare well with observations. The theory was eventually saved by assuming that there are irregularities, or epicycles, that could be calculated to produce the observed planet trajectories. However, as these complementing epicycles were not supported by any evidence other than the fact that they saved the Ptolemaic theory, it is agreed that the empirical observations refuted the theory.

our brain probes, at this point in time, to predict our decisions (i.e., the outcomes of the processes in our brain). In this sense, we are living in dramatic times because with AI and Big Data we are about to get a quantitative answer through the measurement of P. Indeed, the most profound experiment ever done on humans has already been set in motion.

3. THE SOUL

Some will claim that all our arguments so far are immaterial, as we have been considering only the physical aspect of life, while there is a completely different aspect which on the one hand affects us and on the other obeys very different rules. They are of course talking about the metaphysical notion of the soul, which has received countless philosophical or spiritual representations within the mind-body problem, such as life force or life energy.

I once took a taxi in New Delhi, and it seemed like the driver really didn't mind dying. It seemed like he believed so strongly not only in the reincarnation of the soul but also that in the next life he would clearly be rich that his driving appeared like it was intended to be short-lived. I asked him to stop, and I got out. If people are so convinced that the soul is real, it indeed deserves not to be brushed off.

The soul, or as Gilbert Ryle referred to it in defiance, the ghost in the machine, or as popular Chinese culture coined it, ancestor spirits, has been a part of human culture and religion forever. It was a central part of the cultures of ancient Egypt and Greece, and it may be found in ancient Assyrian and Babylonian religions. It of course also appears in the beginning of the Bible (Genesis 2:7): "The LORD God formed the man from the dust of the ground and breathed into his nostrils the breath of life, and the man became a living being."

Philosophers have always been intrigued by the mind-body problem, and the soul may be taken to represent the part of the mind that may not be called body. Socrates, Plato, and Aristotle all discussed the soul. Kant identified the soul as the "I" in the strictest sense. Just like with free will, already in ancient times they understood that there is tension between our belief that everything must obey the laws of nature and our wish to have a soul. They tried to find all kinds of pathways toward understanding. For example, thousands of years ago, the philosopher Lucretius, who was part of the atomistic movement, claimed that, like the mind, the soul must also emerge from an arrangement of particles. In the seventeenth century, Pierre Gassendi, a French philosopher, Catholic priest, astronomer, and mathematician who was influenced by Lucretius, reintroduced atomism into science, but he excluded the soul from the discussion so that atomism became more acceptable to Christianity. Unfortunately, it is hard to imagine that philosophy will ever be able to award us with some proof that the soul exists or does not, and if it does, what its features are.

The soul is not just a great argument against physical determinism. Aside from God himself, and heaven, the soul is also probably, as we note when we discuss the theory of the denial of death, the grandest construction of symbolic immortality (see Part III).

The evolution of the idea of the soul probably followed a fascinating multipronged path. It could have started with people trying to understand what the difference is between the human body a moment before death and a moment after. They called this difference, this elusive force of life that they could not see, the spirit or the soul. In parallel, it was connected to religious stories concerning the creation of humans by God, or the existence of heaven and the afterlife. Perhaps it was also used by rulers to calm the masses when there was unrest due to inequality or when facing death in battle or due to pandemics.

In the language of the theory of the denial of death and its claim that our culture is nothing but terror management—terror of the understanding that we all eventually die—one may also speculate that the notion of the soul was enthusiastically promoted by creative thinkers, or terrified thinkers, or both, who had to do some effective terror management aimed at making available some denial of death. In any case, it is evident that humans welcomed the concept of the soul, as it served well their need for some sort of immortality.

But perhaps it goes even deeper than that. Perhaps some of the promoters of the idea of the soul were thinking long before us about the concept of machine death. Namely, somewhere along the evolution of the idea of the soul, they were not thinking of a tool to outdo the inevitable biological death that would come when they were old—that is, they were not looking to invent something that is immortal—but rather, like us, they realized that they might already be somehow dead, and they devised a weapon of doom, not to bring doom but to fight doom. By that I mean that these thinkers suspected that our physical body, including our brain, is a machine, and to escape this doom which we have coined machine death, these inventors came up with a free entity of transcendence beyond the machine. Indeed, as I have shown previously concerning free will, the need for some mental freedom is ancient.

—⁂—

What evidence do people present in favor of the existence of a soul? Well, there are stories of those who under hypnosis start talking in ancient languages—I believe they call it xenoglossy. Unfortunately, while the observation of the effect of hypnosis seems to be quite well established in science, as far as I know, there is nothing that could be called scientific evidence of xenoglossy.

Another typical argument regards all those who say that after they were clinically dead, they saw a tunnel with light at the end. One may guess that, just like the hypothesis that a memory we have of a dream is probably due to some physical changes that occurred in the synapses while we were sleeping, so too it may very well be that changes in the synapses occurred while we were clinically dead. These then feel like a real memory.

In 1907, Duncan MacDougall conducted an experiment measuring weight loss at death, concluding that the soul weighs 21.3 grams (a 2003 American film used this as its title). Numerous scientists and psychologists later criticized the experiment as not meeting scientific standards.

Science has obviously described the soul as a figment of our imagination. Evidence from brain imaging indicates that all processes of the mind have physical correlates in brain function.[10] However, such correlational studies cannot determine whether neural activity plays a causal role in the occurrence of these cognitive processes (correlation does not imply causation; namely, if things typically appear at the same time, it still does not mean that one of them causes the other, as they could both be emanating from the same origin).

Physicist Sean M. Carroll wrote that the idea of a soul is incompatible with *quantum field theory (QFT)*. He writes that for the soul to exist, "not only is new physics required, but dramatically new physics. Within QFT, there can't be a new collection of 'spirit particles' and 'spirit forces' that interact with our regular atoms, because we would have detected them in existing experiments." He concludes by saying: "There's no reason to be agnostic about ideas that are dramatically incompatible with everything we know about modern science."[11]

In the book *The Soul Fallacy*, Julien Musolino, a cognitive scientist, states that the mind is merely a complex machine that operates on the same physical laws as all other objects in the universe, and that currently there is no scientific evidence whatsoever to support the existence of the soul. He even claims that there is considerable evidence that seems to indicate that souls do not exist.[12]

We evidently still know so little of what there is to know that personally I would be careful in stating that we have proof that the soul does not exist. In my opinion, proof that something exists is much stronger than proof that it does not. Or as John Bell, a famous physics theoretician, put it, proofs of what is impossible often demonstrate little more than their authors' own lack of imagination.* The only thing that is clear is that we have not been able to measure it.

Can we ever really prove there is no soul? Well, some futurists, like Ray Kurzweil, think that pretty soon computers will be complex enough that the whole human experience could be transferred into them, and in this sense, we will live forever as the same people we were but from within a silicon-chip computer. If indeed we will be the same people exactly, then there was no soul, as everything would now be a result of well-known physical processes in the computer, unless someone wants to assume that the soul is still connected to the individual even after they have been uploaded to a mechanical device.

The above is somewhat similar to the teleportation test, as it is presented in *Star Trek*, where they beam people from the spaceship to the surface of a

---

*One of many examples of an error in judging what does not exist is the 1651 book by an Italian Jesuit, Giovanni Battista Riccioli, titled *Almagestum Novum*, in which he gave seventy-seven proofs of the impossibility of the Earth's motion, in particular its rotation. Riccioli's book was written against the Copernicans, and he was no fool. In one historian's view, Riccioli produced "the lengthiest, most penetrating, and authoritative analysis of the question of Earth's mobility or immobility made by any author of the sixteenth and seventeenth centuries." Scientists, even great scientists, quite often make mistakes in their predictions of what is impossible. For example, the great Lord Kelvin stated in 1902: "Neither the balloon, nor the aeroplane, nor the gliding machine will be a practical success." It is also claimed that he said just before the big revolutions of quantum and relativity, "The future truths of physical science are to be looked for in the sixth place of decimals" (namely, no major novelties will be discovered), though some contest the fact that he indeed said these words. Another example is due to Albert Michelson, from the famous Michelson-Morley experiment on ether, which was thought to be a material filling space, who said in 1894, "It seems probable that most of the grand underlying principles have been firmly established." Some even predict the *End of Science* (by John Logan, 2015). For a review of past (wrong) predictions that nothing new will be found, see A. M. Silverstein, "The End Is Near!": The Phenomenon of the Declaration of Closure in a Discipline, *History of Science* 37 (1999): 407-25.

planet. Because all the transporter machine knows how to do is to reassemble the physical state, if people are the same after transportation, then they were nothing but their physical state. Again, someone may wish to conjecture that the soul follows you even after you have been taken apart and reassembled thousands of kilometers away. By the way, teleportation, namely the complete copy-pasting of a quantum state is already taking place in quantum laboratories. Here, different from the science-fiction version, matter is not transported, but rather what is transported is the information of how to build an exact copy somewhere else. If we ever manage to build a person in this way, it will serve as an interesting test regarding the soul.

Whatever the case may be, there is yet no shred of a hint that anything like a metaphysical soul exists, and that's why we still call it metaphysical. On the other hand, as noted, I believe there is no proof that the soul does not exist. Nevertheless, as we have done previously, I think we can analyze the idea of the soul as an antidote to determinism in general terms and still come to some insightful conclusions.

Let us therefore assume there is indeed some metaphysical thing which we refer to as the soul, something outside our physical state, or simply something that is physical but beyond our current understanding of nature, and outside of what our instruments can sense. And let's, for the sake of argument, say that the soul of a person is a good soul; namely, it always votes for good. Is that freedom? Is that free will?

So, let's assume that the soul has free will. It can even sometimes choose to be bad. And let's consider cause and effect. In a deterministic world, A always leads to B. Will the free soul exposed to A sometimes react by evoking action B and sometimes C? If so, then for the same conditions exactly, for exactly the same situation it is encountering, it will sometimes choose B and sometimes C. Based on what? All it has to go by is situation A; this is all the information at its disposal before it makes a choice. This means that there is some randomness involved, but randomness is not free will, as it is no will at all.

There is of course also the possibility that our soul is simply sometimes irrational, which, as we already noted, is not tantamount to random. Even irrational behavior can be a result of deterministic processes. Irrationality may simply be viewed as a faulty feedback loop.

Be that as it may, we again find ourselves needing an objective test. For example, it could be that while our physical brain is deterministic and predictable, our soul is not and in this sense saves us. The crucial thing to understand about our new observable P is that it does not care what the source of our decisions is; it does not care if there is a soul or not behind our decisions. It just cares about the end result, the human outcome, the decisions and actions taken by a human. This is the ultimate test, and it can all be measured by P, an objective measure.

In case we are found to be predictable by AI and Big Data, we can only conclude that our feeling of having a free agent called a soul, just like our feeling that we have free will, is simply a result of the complexity of the machine, which mimics in the eyes of an external human viewer, and even in our own view of ourself, some form of freedom of choice. There may still be a soul, but in the case of a high value of P, it is predictable and therefore deterministic. This nontrivial machine that is us, together with the obvious wishes of our ego, makes us falsely believe that we humans are adorned with all kinds of wondrous things. In sharp contrast, our new observable P has the power to set the record straight in a way that will shake the very core of our existence.

Part III

RETHINKING THE OBVIOUS IS WORTH FIGHTING FOR

3

The Hard Road to a New Understanding

Dear Eve, by now I know you love philosophy. I just noticed I have actually used *love* twice, as *philo* means love as well, and *philosophy* means love of wisdom. I am not sure you would agree with philosophers such as Plato, Descartes, and Kant, who thought that science is secondary to philosophy, but clearly you think that science without philosophy does not see the big picture. I feel the same. Consequently, I would now like to take a short pause from the scientific discussion (I return to the science of predictability in Part V where I discuss neural network noise and quantum randomness). In this philosophical pause I shall deal with the more speculative task of interpreting the results of the experiment, namely, the predictability of humans, in terms of how we view our life and death. That is, if we assume that a high level of predictability will indeed be found in humans, what are the implications for how we view human life? At the end of the book, in Part VI, I return to this discussion, as well as the debate over a possible pill that may perhaps cure us.

I must take this pause now in the middle of the scientific discussion to emphasize what is at stake. Obviously, in the introduction and in Parts I and II, I have already hinted at what the implications of human predictability are, so you already know the gist of it, but in case you have the time to follow your curiosity to the bitter end, I thought I should describe the main contemplations, arguments, and conclusions in a more detailed and comprehensive manner. This is done in Parts III and IV. Thinking back on that day when you, a smart filmmaker from New York City, magically appeared in my lab, I now have the feeling that you already understood what is at stake, but I clearly did not.

Some may view the effort to redefine life with some objective scale, which goes beyond the standard definition of a functioning heart and general brain activity, as bound to fail, as anything beyond the fundamental biological functionality of the organs is no more than semantics or subjective in nature, where by that they would mean that it has nothing to do with exact sciences or an objective reality, but this is clearly wrong. All modern models of the brain, which aim to quantify brain processes with mathematical precision, will eventually enable us, not to say force us, to accept that different healthy brains, that is, biologically functioning ones, possess different values on the spectrum of brain parameters, parameters that will eventually be associated with levels of consciousness, the latter being the scientific essence of human life (e.g., a rather popular theory, integrated information theory, has a mathematical value for consciousness represented by the Greek letter *phi*; see Appendix B and references therein).

However, interpretations are clearly more subjective, and in this sense more philosophical in nature. So let me now take a pause, which may be seen to be philosophical in nature, and consider a possible interpretation of the measurement of our observable, predictability P, with what concerns human life. Interpretations of experimental results may seem to some to be quite easy, but when it comes to rethinking our very core, they actually become extremely difficult.

But first, let me take a hard look in the mirror and ask whether I am open-minded enough to objectively analyze the question of human predictability and its dramatic consequences. The obvious—that is, what is obvious to us—defines our comfort zone, and thus rethinking the obvious means leaving our comfort zone. This is why Sir George Bernard Shaw said that all great truths begin as blasphemy, and why Charles Darwin, who felt that his new theory forced him to abandon his belief in creation by God, wrote, as noted, that it was as hard as confessing to a murder. Leaving a deeply engraved comfort zone typically requires us to vehemently fight against ourselves, but this freedom is almost always worth fighting for.

If we are to seriously rethink our core, let us start from the very beginning. To reconsider the definitions of human life and death, we must first realize that concerning our views on matters that form the foundation of our mental well-being—especially those that we feel are obvious, and especially when the matter concerns the most precious gift we have been given in this difficult life, which is life itself—our views may be highly biased, as they are tuned to fulfill our needs rather than to expose us to objective truths. Freud and many others have been explaining this to us for a long time. We should therefore expect that when it comes to redefining human life and death, our mind may be strongly entrenched, producing friction whichever way we try to move. It was thus clear to me I had a long way to go.

A measuring instrument in a laboratory is an objective entity. It has no personal agenda. No ego, needs, wishes, or biases. But we, the observers trying to forge a complete picture of reality, are subjective entities full of ego, needs, wishes, and biases. A very simplistic yet insightful example of our subjective mind may be found in the dress photograph that became a viral internet phenomenon in 2015. The dress was thought to have different colors by different people. Even at such a trivial level of identifying colors, our mind may be biased. When Hollywood covers the faces of its actors with appropriate makeup, dresses them in particular clothing, and asks them to walk funny, it's easy for all of us to accept that they are zombies. But what if the difference between us and zombies is much more subtle, much more intricate? Can we become objective enough to examine subtle divisions between life and death? Can we digest a new view of ourselves?

As noted in the introduction, many artists and musicians, as well as technological innovators, take mind-altering drugs to detach their mind from the obvious, as the obvious is a static comfort zone. More so, the obvious is a code name our subconscious sends us every time it knows that our subjective well-being requires something to be true even if it is not. We then say, "Ah, that's obvious," rather than engaging in some factual analysis that we fear may be painful. In the future, if some avant-garde publisher has the money, and the law will allow it, and he/she could also find a way to ensure everyone's safety (e.g., not becoming addicts, not to mention overdosing), perhaps they would be bold enough to attach, in a small plastic bag, a tiny amount of some drug to the cover of each paper copy of books attempting to journey far away from our comfort zone, to enable the mind of the reader to look through the crack presented in such books with a cleansed vision. In contrast, our brain has such extraordinary complexity and capabilities that I believe we can train it so that it is also able to wander away from our comfort zone without drugs, so in the following I attempt to briefly describe the process that helped me rethink my personal obvious (in Parts V and VI, I present some more ideas on enabling the mind to rethink the obvious). If you believe your mind is already able to rethink itself, then you may catapult yourself directly to Part IV. Mine was not.

When you visited my laboratory and asked me if the atoms are alive, I told you I would have to think about it. I had no doubt that it was a loaded question and that the first step in addressing such an elaborate riddle must be to identify and map the possible sources of error, especially those limitations and biases originating from the mere fact that I am a subjective human who is asking a question about himself. There is also the possibility that my common sense will fail me, for we indulge so very much in our common sense, don't we? I recalled Einstein as saying that "common sense is nothing more than a deposit of prejudices laid down in the mind before you reach eighteen." To start the inquiry in earnest, it was thus clear to me that I would have to unthink myself.

Consequently, this part of the book is actually a sort of preparation, intended to briefly describe the techniques, barriers, and pitfalls I was thinking of before diving into the question of the definition of human life itself. I decided to briefly share with you this process I went through, in case you find it interesting and perhaps even helpful.

Finally, in the following I sometimes use the plural *we* and *our*, as is customary in academic documents, but when it comes to limitations and biases, and other negative aspects of the human when he, she, or it tries to think objectively, I am solely referring to my own personal shortcomings.

1. TRUTH AND OBJECTIVITY

With what concerns death and life, and the exact point in which the latter ends and the former begins, we are in an intellectual and emotional minefield. To be objective, we need to heighten our senses to their utmost capabilities, wear armor vests, and become zealots of truth, ready to wage war within ourselves in its name. So let us appreciate what these words, *truth* and *objectivity*, really mean and the transformation we must undergo to aggressively seek them.

Let us begin by acknowledging that truth is very hard for humans, especially when it reflects badly on them. Indeed, Immanuel Kant said that humans always find some selfish reason to twist the truth, to say something that reflects well on them.[1] This is especially so when we come to the topic of our life and our death. To avoid Kant's razor-sharp description of our faulty human software, we undertake an almost impossible Houdini move. We must leave our own body and detach ourselves from our ego, no less. In fact, to really know ourselves, we need to spend so much time outside ourselves that we even run the risk of losing our own self-citizenship. Namely, to genuinely view ourselves from the outside, we need to some extent to reduce our level of loyalty to ourselves, to our personal flag.

If we observe the typical miserable human condition regarding truth, we can perhaps formulate a law of faith to help guide us in our search for the truth. It states that before we believe something is true, we should make damn sure we know well all the reasons our environment and our fears and our ego, and even our love and our hate, have in trying to convince us that this thing is true even when it is not. For example, as I already noted, most people feel they are part of the same religion their parents ascribe to, putting into doubt that these beliefs have anything to do with an objective search for truth.

As I was thinking about the concept of truth, I reminded myself that the starting point of any candid search for the truth is deep skepticism, the kind that allows us to erase the very foundation of our beliefs. If you like, this skepticism is an antidote of sorts. If we want to chase after the truth, like someone chases a hurricane, we must become ardent skeptics. Becoming a real zealot of skepticism, this must be our religion. Of course, skepticism in everything that is currently known may be dangerous and may take us down a deleterious road,* but if we are aware of the dangers, skepticism is by far the most advantageous choice.

I believe Karl Popper once said that our dreams and our hopes should not influence our conclusions, and as we search for truth, the best course of action should be to start by criticizing our most cherished beliefs. He added that to some this may seem to be a distorted plan, but it would not seem so for those who seek truth and are not afraid of it.

---

*One of the immediate consequences of unchecked skepticism are conspiracy theories that may even have the power to undo the adhesion of society. An amusing yet insightful story is that of the Birds Aren't Real movement. See the TED talk at https://go.ted.com/5sLm and an interview from Johns Hopkins University at https://hub.jhu.edu/2024/02/07/birds-arent-real/.

In order to believe unambivalently in something, without any shred of skepticism, the intelligent mind must lie to itself. But even if one does not lie to oneself, a deep form of skepticism is not ensured. The latter may only be born out of the understanding that reality is infinitely complex, and with our limited sensory capabilities and our limited brain power, and especially with the very short time in which we humans are properly investigating this reality, we are surely only beginning to scratch the surface, thin layer after thin layer. Plato's cave allegory shows that already thousands of years ago, some people realized that our understanding of reality is very partial.* Such deep skepticism may then open the door for growth of the mind.

This comprehension of the infinity of reality has many insightful representations. For instance, the totalitarian principle, closely related to an older principle of plentitude, states that everything not forbidden is compulsory.[2] Forbidden means that if we believe in an existing law of nature and there is ample empirical evidence to support it, we cannot go ahead and imagine something that contradicts this law. For example, it seems futile now to try and contradict the basic conservation laws of momentum and energy. But other than that, the ultimate reality most probably goes beyond anything we think we know or can even imagine. The totalitarian principle was used by many, such as the author T. H. White, who tried to describe an ant colony. It was also used by physicist Murray Gell-Mann, who got the Nobel Prize in the 1960s, to describe his theory for the new particles that were being discovered. Some quantum physicists also use it to describe the many-worlds interpretation of quantum mechanics, where when there is some junction of possibilities, every possible occurrence does indeed happen, each in another parallel universe.

A nice description of this infinity was given by a physicist named Lichtenberg, who said that the wise man seeks the truth, and the fool finds it. Indeed,

---

*Indeed, the idea that our limited sensory capabilities and our limited brain power do not allow us to see reality as it is goes back thousands of years, and Plato's cave allegory (or as it is sometimes simply called, the "allegory of the cave") is a good example. It describes how reality is not what it seems to be. In his allegory, people living in the cave can only look toward the back of the cave. On the back wall they see shadows of real objects, and they believe these shadows are the reality, although they know nothing about the real objects. The real objects are between a fire or source of light and the people, but as the people cannot turn their heads, they cannot see the fire or the real objects and they only see the shadows. In fact, the ancient discussion goes on to say that after a long time, if someone freed the people and asked them to look at the fire and the real objects, they would be horrified and run back to their comfort zone.

the wise person understands that what he has found is only the next layer of an infinite number of layers.* I also reminded myself of what William Blake wrote: "If the doors of perception were cleansed, everything would seem to man as it is, infinite." I felt it heartwarming that this is where Jim Morrison found the name for his band (via the book by Aldous Huxley). Finally, there is a quote I am genuinely fond of. I found it in a book by Nobel laureate Tsung-Dao Lee. He quoted the Tao Te Ching from about the sixth century BC, which said something like, "Any principle that can be stated cannot be the absolute principle, and any name that can be given cannot be a permanent name."

Let us now put aside for a brief moment the deep skepticism that we must acquire in order to be able to cleanse our doors of perception so that we can see far away beyond the horizon, and speak of the scientific criterion for truth.

This criterion very clearly specifies how to move closer to what is true and farther away from what is untrue. Beyond the recognition that truth is always a temporary truth, it provides a structure in which hypotheses are made regarding nature concerning what causes what and how systems evolve with time, and then these ideas are empirically tested. A new description of nature becomes a step toward the truth only when proven empirically in several independent observations and judged by independent observers.

It's a great method that is based on cause and effect, namely that everything that happens is simply caused by something that happened earlier. The scientific method is of course also based on the belief in the ability of human logic to understand this causality, and on top of that in the ability of humans to empirically test the human descriptions of this causality.

It indeed seems that science has given humans the best criterion so far for what may be considered to be true or, more accurately, what may be

---

*This belief in the infinite complexity of truth has a wide variety of very practical consequences, even at the level of how we educate the next generation. For example, the big difference between the way of the religious truth seeker and the way of the secular one is that the religious person believes that the truth has been found, and he loves his parents and his teachers and respects them by strictly following in their footsteps and making sure his children do the same, while the secular person loves and honors his parents and teachers by acknowledging their attempts to uncover the truth and by trying to discover a deeper truth himself, and he then hopes that his children as well find a better path closer to the truth than what he had found.

considered to be untrue. I think this is actually the greatest achievement of science, even more important than the actual knowledge about nature that it has provided, as it enabled humans to grow beyond their tendency for superstition and fear of the unknown, and one may even say that this criterion of truth has the philosophical power of an emancipator and that it freed us from considerable slavery of the mind. Science has indeed followed Kant's motto for enlightenment, namely, *Sapere Aude,* "Dare to know," or dare to know things through reason.

Finally, a cautionary statement concerning truth in this book: The text is full of facts but also full of speculations regarding the future. These speculations may be considered learned conjectures or grounded extrapolations, but they are speculations nonetheless. I will try and draw a clear linguistic borderline between what are facts and what are speculations, but in general, the two are simply divided by past and present versus future.

2. FIGHTING OUR EGO

Eve, we are laying down the foundations for a new understanding. But even being a zealot of truth and objectivity is not enough—so I told myself as I was thinking of the methodology that could perhaps allow me to answer our question on what constitutes human life. We must identify our greatest enemy and find ways around it. In this case, our archenemy is the never-ending human endeavor toward the denial of death. As we examine death, this will be the nemesis within us, and the main culprit is our ego.

Humans are probably the only animal that knows it will die, and we are terrified of it. We are the only ones who know we will eventually lose the fight, and it is driving our ego crazy.

Let us start by examining other fronts on which our ego is at war with the facts.

They say that three people have hurt the narcissism or the ego of the human being the most: Charles Darwin, who showed that we are not part of some meaningful plan; Nicolaus Copernicus, who showed that we are not the center of the universe; and Sigmund Freud, who said that we are controlled rather than in control.

From the above three giant thinkers came serious blows to our ego. For example, out of the scientific findings of Copernicus and other astronomers came the Copernican principle, which asserts that we are not special in any mystical sense. We are not part of some godly plan that puts us in the middle. It starts with the assertion that the sun and planets and stars are not revolving around us, and then it transforms into a philosophical proclamation about our lack of uniqueness.

Just to show how hard our ego can fight back even in the face of compelling evidence, let us also briefly recall the story of Darwin and evolution.

Specifically, I would like to ask how two hundred years after Darwin landed in the Galapagos Islands, and after Alfred Russel Wallace arrived at the same conclusions as he collected more than a hundred thousand specimens in Malaysia and Indonesia, and after hundreds and thousands of delegations and studies reinforced their conclusions, at least in principle, people still find it hard to accept? Have we learned nothing from Galileo's trial? What bothers me is not so much the resistance to the theory, which like any theory may fall one day, or be extended to cover new possibilities like epigenetics, just like Einstein's relativity is an extension of Newtonian dynamics.* What bothers me is the emptiness of the criticism, without having any wish to first seriously study the theory and examine the evidence, and without bringing to the table of debate any rational arguments against the overwhelming body of empirical deduction. Such empty criticism means that something other than an objective search for the truth is afoot. I believe this is the ego at work much more than it is a religious bias.

There are countless books in which the huge body of evidence concerning Darwin and his theory is detailed, but I would like to add one personal experience I had. I once visited the office of Professor Yoel Rak at Tel Aviv University, where I was enchanted by the numerous glass cupboards housing many skulls of different shapes and sizes, spanning millions of years, which he and his colleagues had found in the ground. Gazing into those glass cupboards, you could really see evolution at work; you could follow the transition from apes to humans.

---

*By the way, if I would be hard-pressed to present some argument in favor of some "intelligent design" theory, and against any kind of evolution or some fast adaptation, I think I would note the four pairs of holes that are positioned in the pelvis at the bottom of the spine. If I am not mistaken, some nerves go through these holes. The holes really look as if some engineer thought of them. I decided to include this rather strange footnote, to hint at how easy it is to be attracted to unfounded theories. Please see footnote a few pages ago about the Birds Aren't Real movement.

So many people, even nonreligious people, do not believe in our Darwinian origins, even in the face of unambiguous evidence, that indeed the conclusion must be that a very strong internal force is at work. You and I may say, "Well, I do believe in evolution, so my ego is not clouding my judgment," but in fact Darwin is only the tip of the iceberg in how the ego toys with us. Let us therefore go to a more serious threat to the ego: death itself. We might again think that our ego will not be able to hijack our conclusions regarding death, but this is our ego speaking, and it is in fact acting as a double agent.

Schopenhauer stated that the fear of death is the beginning of philosophy. But it is in fact much more powerful than that. Nobel laureate Saul Bellow said that human culture is nothing but an ingenious device intended to make us forget that in the end we will surely die. I guess his premise was that civilization is geared to suppress our fear of our mortality, where examples of how civilization and its culture go about doing this are to be found in the concepts of God, the soul, heaven, and of course the Messiah, who will bring everyone back to life. Religious beliefs indeed offer flamboyant structures of immortality.

If we are so terrified of death, how will we be able to objectively digest the idea that to some extent we are already dead, as brought about by the concept of machine death that we are considering here?

Let us thus further consider this barrier of thought. In 1973 a more detailed theory regarding the human effort to suppress the fear of our mortality was published by Ernest Becker, adding to the above tools of immortality, engineered so far mainly by religion, also secular tools including the community of peers, children, and even flags. It was named the theory of the denial of death. Not that it proves anything, but in 1974 Becker received the Pulitzer Prize for his analysis.

In the years that followed, numerous other people continued to work on this theory, like Jeff Greenberg, Sheldon Solomon, and Tom Pyszczynski. They called it terror management. They essentially added to, or even replaced, the main driving forces behind our actions à la Freud, such as sex, with a new kind of force aimed at denying death or, in other words, cheating it.[3] I don't believe we should discard sex as a strong impetus, but it may well be that these two forces work in concert.

I once heard a philosopher, Sam Keen, say that death is an insult to our spirit. I now think he was severely understating the matter at hand. It's much worse than an insult. I think these guys were right when they talked about terror.

They started speaking of symbolic immortality. If someone has special skills, his symbolic immortality would be encoded in some meaningful impact like a great book or movie or invention, or generosity, or even some heroic death. These would live on in the minds of people and in texts long after he or she is gone. If someone doesn't have exceptional skills, their symbolic immortality comes from their belonging to something larger than themselves, something that would outlive them, such as some ideology, some collective, like being loyal to some religious or national flag.

We may say regarding the above idea of meaningful impact, "But something meaningful in one culture is not meaningful in another, and this weakens our immortality." Yes, and that's why we work so hard to convince ourselves that our culture is superior, whether it is in sports, science, the economy, on the battlefield, or even spiritually. If we have a meaningful impact in a superior culture, our impact is meaningful in a universal way. The same goes for the above-noted need to belong to some unique collective. It gives much more immortality if the collective we belong to is superior, as superiority must indicate that this is the true universal way. Perhaps that's why very religious or very ideological cultures brought so much intolerance, so much violence. I guess we should not be surprised if eventually they will find out that these kinds of mental processes also stand at the very core of racism.

Like for Freud, here also there is a big difference between the conscious and the subconscious. While, in order to achieve immortality, our conscious mind may invent things like God, the soul, and the afterlife, the subconscious may be much more destructive, as it's making us do things we don't really understand.

Denial-of-death theorists have done quite a number of experiments to see if reminding people of death makes them cling more tightly to their culture. That is, would such reminders lead people to be harsher toward someone who violated the morals of their culture, or make people more reluctant to treat the symbols of their culture inappropriately?

For example, they took two groups of judges and asked them to fill out some personality questionnaires. For one group, these forms included asking them what emotions their own death arouses in them and what they think will happen to them when they die. They then asked the judges in both groups to decide on the punishment for a case of prostitution. The judges had all encountered similar cases numerous times in the past. The judges in the group that had mentions of death in their questionnaires gave average fines of $450, while for the other group the average was $50.

The researchers also took two groups of students. Again, they had to fill in some forms. As before, only one group had mentions of death in the form. Later on, each student had to perform a task on his or her own. For example, they had to separate some liquid ink from sand that was in it. They had all kinds of unusable stuff available to them, and the only thing they could actually use to filter out the sand was an American flag. The group with the mentions of death took much longer to utilize the flag, knowing it would be stained, and felt tense after doing so. Similarly, they had to put a nail in the wall, but they had no hammer. All they had were all kinds of unusable things like cloth, and the only thing they had that they could use to hit the nail was a Christian cross. The group that was exposed earlier to mentions of death took much longer to use the cross to complete the mission.

So there seems to be a direct link between our thoughts of death and our clinging to our collective culture. As claimed by the above scientists, I believe this indeed hints that we are using our cultural membership as a tool for symbolic immortality. The bottom line here is that our subconscious and our ego seem to be engaged in terror management concerning the topic of death, and this clearly requires caution on our part when we try to think objectively about death.

In the context of the ideas of the theory of the denial of death, let me add that, in pondering about Copernicus, Darwin, and Freud and the blow they inflicted on our ego, if we think about it carefully, their theories and findings did not only hurt our ego directly; they also hurt our efforts toward symbolic immortality, and this probably hurts our ego the most.

As I noted, humans are likely the only animal that knows it will die, and we are terrified of it. We are the only ones who know we will eventually lose the fight, and it's driving us crazy. If fear of the eventual physical death drives us crazy to such an extent, imagine what it will do to us if people knew that the present-day elucidation of death is wrong and that the definition of human life and death and the border between them on which we are so well rehearsed, namely that life stops and death begins simply when the heart or brain stops functioning, represents only a small part of actual reality. And that many of us, if not all of us, are already to some extent dead, although we think we are alive. Although we are speaking here of mental death, we use the word *death* not as some metaphor, as mental death is real death. If we open ourselves to such a thought, it is a Pandora's box, and so many mental structures that keep us safe will collapse.

One may thus hypothesize that our present definition of human life and death is a shield that protects us. It is a necessary condition for our existence, for our mental survival. It is this necessity that stands at the foundation of this common definition of when life begins and when it ends, not its veracity. This very rudimentary, even primitive, definition that we are trained to use is in fact the ultimate denial of death, as it vehemently denies the concept of machine death, or any other more subtle definition of death for that matter. In this sense, Becker and his theory only went halfway, as he did not criticize the definition itself. Taking into account the theory of the denial of death and the role played by the present-day definition of death as a protective shield, it is safe to assume that it will be very hard for us to accept a new definition.

But if we are about to fight against the denial of death, which originates in the ego, we should not be trying to erase the ego. This would be somewhat like trying to push the impossibility button (a concept appearing in the *Hitchhiker's Guide to the Galaxy*), as the ego is an important part of us, and without it we are not we. On the contrary, we need a huge ego, but of a different kind. To reach total skepticism, the ego needs to be so tremendous that it allows itself to aim at a target that is against its own interests, because self-preservation, which is the root of all our self-lies, seems to it to be too mundane, unworthy of its respect. Namely, as we already noted, a person needs to be totally loyal to something else, something outside of himself or herself. This should not be confused with self-hatred. This is all about pure and relentless pursuit of truth. "Can my ego make the necessary transformation?" I could not stop asking myself this as I attempted to design a methodology with which I could examine our question of what human life is.

3. WORDS ARE NOT ENOUGH

We continue with our efforts to prepare ourselves to become objective thinkers, able to analyze our life and death without bias. Deciding on a new, earlier point at which we will be pronounced dead is in a sense like deciding when we will die, which is in a sense like suicide. It was clear to me that this is going to be very hard. Before addressing the question of human life and human death head-on, I tried to find additional limitations and barriers to my ability to think about this problem.

We all know about the limitations of our senses and how they hinder our ability to fully understand reality. For example, our eyes can only see a small fraction of the electromagnetic spectrum. We can only see wavelengths from about 380 to 750 nanometers. Imagine what we could know about reality if we could also see X-ray, UV, or infrared light. Similarly, our ears can only hear frequencies between 20 and 20,000 hertz (cycles per second). Imagine what we would know if we could hear more things. People are generally aware of these limitations (limitations that some people dream of fixing with the advent of cyborgs), but few are aware of the hindering effect of language. In fact, the crippling effect of language goes far beyond the passive shortcomings of the senses. It seems language plays an active disruptive role in how we think internally and how we think collectively.

Language brings with it the bias of previous notions, expectations, or common sense. This bias is usually so subtle that it is hard for us to even notice it. Wittgenstein correctly noted that "the boundary of my language is the boundary of my world." Indeed, Thomas Kuhn, who studied philosophy of science, fantasized a pure language that would enable objective observations. Unfortunately, he ended up being a pessimist concerning the possibility that an objective language allowing us to really see would ever be found.

There is a well-known Zen story about two teachers, Kabir and Farid, who met for two days and did not exchange a word. To their amazed followers, they said, "When silence can speak, what is the need of language?" Eve, like these two teachers, I wish we could speak in silence, but alas ordinary people like me need language to convey thought, and language is very faulty, to the point of being unworthy of our trust.

—ɯ—

In my attempts to understand the limitations imposed on me by language, I found that artists also noticed how limiting language is, to the point that some of them declared language as a threat to humans. Laurie Anderson stated that language is a virus! This intricate thought originates from the postmodern writer William S. Burroughs, who was part of the beat generation attempting to rethink the standard narratives of our culture. One may take this disappointment in language further and state, as has already been done, that although we think we invented language for our own good, it has outlived its usefulness and in fact is holding us back. Some have taken this trepidation to the extreme and claim that language is actually a parasite that infects people and feeds on them. It's not even a symbiotic relationship; language is the hunter, and we are the prey. Because evolution is aimed at the survival of the network and not its individuals, it may very well be that language did not evolve to help the individual think but rather to help the collective survive. In such a collective, it stands to reason that language doesn't give us intellectual freedom; on the contrary, it takes it away. In this sense of freedom, one may even wish to be theatrical, declaring that language is a colonial power, as it aims to spread and control. Some philosophers went as far as calling language a fascist agent. But let us leave the above thought-provoking philosophical and artistic expressions and turn to science for actual examples of how language leaves much to be desired in terms of objective thought.

Let me briefly provide some details on how science readily provides tutorials concerning just how faulty and limiting language is. Typically, ocean waves arrive from afar in parallel straight lines, or, as they are professionally known, straight wavefronts, until sometimes they hit an opening between two wave breakers. At that point, the wave, which manages to pass the opening, suddenly has a round wavefront that propagates in all directions. It even manages to go into the areas behind the breakers. This is called diffraction, and it is one of the clear signatures of a wave. Light also behaves as a wave, exhibiting clear signatures such as diffraction or interference. It is therefore natural that science viewed light as a wave. The linguistic pitfalls started to appear in the beginning of the twentieth century in the form of logical contradictions, or oxymorons, and one of them involved the photoelectric effect. The problem is that in the photoelectric effect, light behaves as particles, not waves. This was so important that it is in fact what Einstein received the Nobel Prize for.

In order to understand this linguistic downfall, let me try and present in a simplified manner the photoelectric effect, keeping in mind Einstein's proverb that it's good to explain things as simply as possible, but not simpler than that. (Eve, let me again excuse you if you do not fancy the detailed physics and ask you to go, without any scruples, directly to the third paragraph before the end of this section, starting with "I hope that by now . . .").

The throwing out of an electron from the atom is called ionization, and the energy required to do so is obviously called the ionization energy. If we take a piece of metal and put it next to a lamp, atoms on the surface of the metal facing the lamp will get ionized, and electrons will be ejected from the metal. Now, when we treat light as a wave, we realize that its energy must be spread equally or homogeneously across the surface of the metal piece. As we know the amount of energy coming out of the lamp and we know the size of each atom, we can easily calculate how much energy an atom on the metal surface gets every second. This simple calculation tells us that it will take a relatively long time before we see any free electrons coming out of the metal. However, in reality, we see free electrons immediately once we turn on the light. Einstein said that we have no choice but to see light as particles that hold a lot of energy and that hit only some of the atoms of the metal with enough energy to ionize them immediately. They called these particles of light photons.

But what about the wave properties of light we discussed before? Aren't waves and particles the exact opposite?

Indeed, they are opposite, Dr. Jekyll and Mr. Hyde opposite, and when scientists started to understand that our language does not have an appropriate word for what they were observing, they started talking about duality, namely that a certain reality may have different words, even opposing words, attached to it. Specifically, they started to talk about the wave-particle duality. Differently said, words started to have a loose meaning, or a much more complex meaning.

This particle-like behavior of light was observed in other cases as well. In 1922, Arthur Compton showed that when photons collide with free electrons, they behave just like billiard balls when they collide. This means that photons have momentum just like any other particle, although they don't have mass.*

If you are wondering, then yes, it indeed means that shooting photons from a flashlight gives recoil like when you shoot a gun. So light can exert a force, and it can push you when it hits you. In fact, the 2018 Nobel laureate Arthur Ashkin managed to hold in the air a small glass ball just by shining light on it from below. Johannes Kepler noticed already in the seventeenth century that the tails of comets are always pointing opposite to the direction of the sun from the comet, irrespective of where the comet was heading, and he realized that it must be a force exerted by the light from the sun that pushes the tail, which is made of material ejected from the comet. He then proposed that future spaceships might have a sail, just like sailing boats, and that they would be pushed by the light coming from the sun. For the sake of accuracy, let me add that today we know that this solar wind is also made of charged particles that actually do have mass, and it is these charged particles that probably eliminated the Martian atmosphere and oceans, but that's a different story.

One may ask, does duality work the other way around as well? What about bona fide particles like the electron or an atom? Indeed, in 1923 Nobel laureate Louis de Broglie, who had survived the horrible carnage of the First World War as a radio technician on top of the Eiffel Tower, suggested that all particles are also waves, and he said that their wavelength is simply proportional to the inverse of their momentum. Namely, he found a simple relation between their particle property, momentum, and their wave property, wavelength.

The proof came quite quickly. In 1927, G. P. Thomson (the son of J. J. Thomson who discovered the electron) aimed a beam of electrons at a thin metal foil. The atoms in the foil may be viewed as the wave breakers in the sea, and the gap between the atoms as the opening left between the breakers. When he photographed the electrons as they were coming out on the other side of the foil, he could show that they underwent diffraction and interfer-

---

*The concentration of light into photon particles is sometimes also referred to as the quantization of light or the electromagnetic field. A beautiful example of this (and perhaps a stronger proof than provided by the photoelectric effect) is photon antibunching, which means that when you observe an atom that undergoes an internal decay (i.e., from an excited electron state to a lower electron state), you only observe one photon at a time. See H. J. Kimble et al., "Photon Antibunching in Resonance Fluorescence," *Physical Review Letters* 39 (1977): 691.

ence, just like water waves do, so he proved that electron particles are also waves and that de Broglie was right.

I hope that by now, like me, you are frustrated, perhaps even a bit bewildered, but that's good. Understanding that our language is so limited, so poor, and that consequently our brains are shackled with big fat chains is a great liberator of the mind. I spent quite a lot of time looking at these chains from all sides, examining the craftsmanship that made them so strong, before I felt ready to address the difficult problem at hand.

To conclude this chapter, we may state that we have tried to very briefly sketch the hard labor of objective thought. We have taken a first quick glance at the confines of our perception, and this in itself is a potent antidote of sorts, allowing us to look deeper into the interior of the brain that is us.

The famous surrealistic painter René Magritte stated that he was trying to think as if no one had thought before him. With what concerns a new definition for human life, this is exactly what we must do.

4

Enabling Our Mind to Make a Leap of Thought

1. THE THOUGHT EXPERIMENT

Every time we analyze a problem, we are engaging in a thought experiment that takes our mind and catapults it into new pathways. Our mind is then forced to reevaluate things. As I thought about the question of how to define human life and what I would have to do in order to truly comprehend the problem, it became evident to me that the journey will not simply require the usual thought experiment but rather a thought experiment on steroids, as it will ask me to dissect my own brain. Even worse, it will ask me to reevaluate myself in such an extreme way that it may even bring me to deduce my own death. This is so because if I manage to arrive at a new definition for human life, I myself may not meet the requirements set by this new definition. The new criterion for life may be above and beyond my own capabilities. How can one think straight under the threat of such bad news?

As I was thinking of what methodology I could use, I started by asking myself what great thought experiment I could try and learn from. The answer was Albert Einstein. Let us therefore more precisely delve into the idea of the thought experiment by briefly surveying one of the greatest thought experiments of all time.

Einstein showed that just by putting hints and ideas together in our mind, while having no available experimental data or empirical proof, and by simulating in our mind how the different ideas work together consistently, one could arrive at actual truths regarding our world. (Eve, to convey what exactly a thought experiment is and just how powerful it can be, I now intend to write a couple of pages with somewhat more detailed physics, so if you don't have the patience for it, I urge you hop to the last paragraph before the next section, Section 2, begins.)

Before I give an actual description of the thought experiment Einstein conducted in his head, I would like to briefly describe the conclusions he arrived at and the reaction of the scientific community. It all started when Newton thought there was something like absolute space, namely some absolute reference frame that is more fundamental than all other reference frames. Einstein (as well as Leibniz, Mach, and others) claimed that this cannot be, as how can something affect others but not be affected by them?

Long before there were any experimental observations, Einstein made a leap inside his head. For example, he came to the conclusion that the speed of light should be a constant in any frame—namely, it does not matter if the measuring device is on Earth or on a fast-moving spaceship—and also that this speed is the ultimate speed and nothing can go faster. If this is true, all previous physical intuitions had to be changed.

He was right. For instance, when two similar stars circle each other around a common center, it's called a binary star. At a certain moment, one is moving toward our telescope and one away from it. So, if the speed of light adheres to the same rules as a stone thrown from a moving train, light should be faster coming from the star moving toward us, and the picture of this star should get to us sooner. We would then see a warped image of the binary system, but we don't. An example for the second idea Einstein had, namely that the speed of light is the ultimate speed, may be found in particle accelerators, in which it is known that no matter how much energy we give a particle, it will never go faster than the speed of light.

But Einstein went further. If we believe, so he argued, that the speed of light is always the same, then if one tries to be consistent, one finds that other things must become relative, depending on where the observer is. For example, mass grows with speed, and length contracts with speed, but perhaps the most mind-blowing outcome is that the simultaneity of two events is relative and depends on the frame of reference of the observer. Two people, one on Earth and one in a fast spaceship, see two explosions in space, and one would say that they happened at the same time while the other would say that one happened before the other. And both statements would be right.

The amazing thing is that this relativity of time is not some odd problem with mathematics or observations. Any clock we build, even the aging of our body, would follow these rules, so that if two twins are separated and one goes away in a fast spaceship and then returns to Earth after what seems to him to be a year, his brother may have already been dead for hundreds of years depending on how fast the spaceship was traveling.

Again, he was right. We already observe in particle accelerators that at high speeds, unstable particles live much longer before they decay. But the most beautiful demonstration was in 1971, when two atomic clocks were separated and flown around Earth. One was flown west, and as the Earth turns toward the east, this clock was almost at rest relative to some stable frame of reference outside Earth, which is sometimes called an inertial frame. The other one was flown eastward and was thus moving much faster relative to the same frame.

When the two clocks were brought back together, they had a difference of about three hundred nanoseconds, as expected by Einstein's theory of relativity. The GPS system that now enables all navigation around the world has to take this into account because of the speed of the satellites.

It is perhaps worth noting that some philosophers were wrong to conclude from Einstein's theory that there is no one objective reality. Einstein actually even considered calling his theory the theory of invariance, which means it remains unchanged for all observers, because in his theory, the laws of physics, namely the equations, are the same for all frames of reference, and for him the equations are what describe nature more than the values of the parameters like length or mass or time.

It should also be noted that Einstein's thought experiment took him so far away from the human comfort zone that many objected to his conclusions. I am making a point of presenting these strong reactions by the scientific community, as I expect that any fundamentally new definition of human life will attract similar criticism, even within ourselves, and this has to be factored in.*

Once Einstein came out with his theory, physicists such as Ernst Gehrcke and

---

*To make sure there is no ambiguity, I must clarify that I am in no way whatsoever trying to even hint at the possibility that the theory of machine death presented in this book comes anywhere close to the absolute genius of Einstein and his theories. Einstein is presented here only as a role model I can learn from.

Philipp Lenard found it easy to cast doubt on his complex mathematics. The Nobel Committee was so divided on the correctness of the theory that as a compromise they gave Einstein the Nobel Prize in 1921 for something completely different, the photoelectric effect. The criticism kept on coming. In 1931 a book titled *One Hundred Authors Against Einstein* appeared.* The criticism continued into the second half of the twentieth century. Einstein's theory of relativity is so uncanny that in 1963 a prominent physicist by the name of Herbert Dingle, the former president of the Royal Astronomical Society, decided he is absolutely sure that Einstein is wrong. He wrote to the prestigious journal *Nature*, stating that Einstein's theory of relativity is incorrect and that there is an international conspiracy to try and silence this point of view. Einstein died in 1955, so he could not reply. As a reply to Dingle, and in defense of Einstein, a mathematician named John Synge attacked Dingle by using the artistic rhetoric for which the British are notorious. He addressed Dingle directly by writing something like this: "What if Dingle is pulling the leg of the world? It is to me the most reasonable hypothesis to explain what is otherwise inexplicable. Knowing you as well as I do, I cannot bring myself to believe that you are as stupid as you make yourself out to be. If my hypothesis is correct, I salute you for your sense of humor. No harm has been done. Printers have had a good employment. My humiliation in having been taken in, is swallowed up in my admiration at the way you put the thing across." Another example is Maurice Allais, a Nobel laureate in economics, who said he has disproven Einstein's theory. Indeed, even scientists find it hard to leave their comfort zone. Building on the words of Sir George Bernard Shaw that I mentioned earlier, one may say that every great theory starts as blasphemy in the eyes of those who truly believe in the old paradigm and is destined to end its life as a mere prejudice.†

To really bestow upon us the power of the thought experiment, let us complete the story of Einstein. Please bear with me. The details of Einstein's theories of special and general relativity and the experiments proving them to be correct may be found in many popular books such as that by Will

---

*Available in English on Amazon. See also interesting analysis of this book in the article by a professor of physics at the University of Texas, Manfred Cuntz, "100 Authors against Einstein: A Look in the Rearview Mirror," *Skeptical Inquirer* 44 (2020).

†Another quote I like that describes how hard it is for people to change their mind is by one of the fathers of quantum mechanics, Max Planck, who is often paraphrased as saying that science advances one funeral at a time. What he meant seems to be: A new scientific truth does not triumph by convincing its opponents and making them see the light, but rather because its opponents eventually die and a new generation grows up that is familiar with it.

Clifford, *Was Einstein Right?* For example, the two clocks that were flown around the world in opposite directions actually measured two different effects coming from two different aspects of Einstein's work: The first was the one we discussed, which is related to the special theory of relativity and the fact that as a clock moves faster, the time as it is measured by the clock will tick slower. The second effect is related to the general theory of relativity developed by Einstein ten years later, which describes not only systems with a constant relative speed, but also systems with relative accelerations, namely, when their relative velocity changes over time. Such accelerations also describe free-falling objects above stars and planets, and so the theory in fact also accounts for gravity. Einstein predicted that as a clock is positioned farther away from Earth, it will tick faster. This means that time will pass more quickly far away from Earth than on Earth, and similarly above any large mass. A combination of these two effects, relative velocity and gravity, worked against each other when the clock was flown eastward (as Earth is rotating eastward, the clock has a high velocity, causing a slower tick rate, but the altitude of the flight brings about a faster tick rate) and in unison when the clock was flown westward. This gives rise to the difference that was indeed measured between the two clocks once they were brought together again, namely, about three hundred nanoseconds. This effect of time changing with height due to gravity was given the name *red shift*.

Einstein called this experiment he did in his head a *gedanken* experiment, which in German simply means a thought experiment. In the next section, Section 4.2, we will try to do something similar in order to simulate us, namely, our brain. With this thought experiment in mind, we will then, in Part IV, analyze the meaning of the word *life* in relation to the predictability experiment and the observable P it measures.

And now, to Einstein's thought experiment itself. As noted above, part of the experiment he did inside his head was to understand the effect of gravity. Einstein imagines a very tall tower built on Earth. At the top of the tower, he places a radio transmitter, namely, a source of light, and at the bottom a detector or receiver. Next to the top of the tower, Einstein places a transparent box, say, made of glass, and in this box he puts a scientist, an observer. At some moment, the transmitter on the top of the tower releases a radio signal toward the receiver positioned on the Earth's surface. The transmitted radio signal includes a beep every second, and the observer, who is stationary relative to the transmitter, indeed measures the beep frequency of once per second. At the very same moment the signal is transmitted, the glass box is released so that it starts free-falling under the pull of gravity toward Earth alongside the tower. The observer inside the box sees the radio signal moving toward Earth but at the same time sees the receiver on Earth speeding toward him. As he knows about the Doppler effect, which makes our ear hear a higher pitch or frequency of a siren when a police car or ambulance is moving toward us, he concludes that the receiver on Earth will recognize the beeps in the signal as coming at an interval of less than a second, namely, at a higher frequency. As frequency and time are essentially the same thing, the person on Earth would conclude that the clock on the top of the tower ticks faster. This even means that people living on the top of the tower would age faster.

The flight of the two clocks was not the only experiment that confirmed Einstein's thought experiment. In the 1970s, instead of building the tower, they fired a rocket. It was sixty years after Einstein imagined it! The very accurate clock in the rocket made it to a height of ten thousand kilometers, and it proved again that Einstein was right and that a thought experiment is a very powerful tool. But I think the most beautiful proof came in 1919 when, during an eclipse, Sir Arthur Eddington was able to observe the bending of light by the sun, light from faraway stars that had to pass near the sun on its way to Earth. He saw the bending by observing that suddenly the well-known position of these stars in the sky shifted when the sun was about to come in between them and Earth. The details of the orbit of planet Mercury gave yet another verification, I believe it was the first.

The prediction of this bending of light was also the result of Einstein's thought experiments. He again used his transparent glass boxes or elevators; this time he used many of them, one for each region of curved space-time, whereby the light from a distant star, on its way to Earth, goes through one elevator after another. He imagined a multitude of glass boxes with people inside falling toward the sun, with light bending in each of them due to the equivalence principle. He evidently achieved a hell of a lot with his free-falling transparent boxes.

Einstein was of course not the first to think about gravity. There was Newton, as well as many others less familiar, such as Henry Cavendish who set out in 1797 to weigh the world, but Einstein did the ultimate thought experiment, with mind-blowing consequences.

—⁂—

But speaking of the complex reality of us humans and our life, which is what we want to examine, can the method of a thought experiment, or simulation, also somehow be helpful without the mathematics Einstein used, mathematics that enabled him to check for consistency and arrive at quantitative conclusions? This remains to be seen, but philosophers would perhaps say yes, as they typically analyze the world in their head without any mathematics. More so, Einstein had a clear physical system in mind, whereas we want to simulate us. We are not a simple well-defined system by any stretch of the imagination. As we are complex entities, we will need the simulation we do inside our brain, namely the thought experiment, to be to some extent all encompassing in the sense that it will have to take into account many effects and parameters, or as we in science call it, many degrees of freedom.*

2. OUR BRAIN IN A GLASS VESSEL

We now plunge into our thought experiment, and let us first use it to get rid of the irrelevant noise, namely, to deflect any unwanted disturbance. In this examination of us, namely, of who and what we are, we are our brain, and consequently the body outside the brain is irrelevant, and we need to dispose of it. Let us therefore imagine a table in the center of a laboratory. On the table

---

*We need to clarify that there is no claim being made in this book that we understand how the brain works. We will be claiming that one may come to interesting insights about the brain from general considerations about the physical nature of the universe, especially in relation to the predictability experiment measuring the observable P. This is the thought experiment, or simulation, we would like to undertake.

PS: Science is just scratching the surface of understanding the brain. Thousands of scientists around the world are working hard, with numerous experimental tools and on several different levels, trying to understand how a network of neurons gives us cognition. Beautiful insights have been gained. One of my favorites is, "The brain is in the game of optimizing neuronal dynamics and connectivity to maximize the evidence for its model of the world" (Karl Friston, "Does Predictive Coding Have a Future?," *Nature Neuroscience* 21 [2020]: 1019–21). To get just a tiny taste of recent work, see, for example, the January 14, 2020, report of Jordana Cepelewicz in *Quanta* magazine titled "Hidden Computational Power Found in Arms of Neurons," where she reports on Albert Gidon et al., "Dendritic Action Potentials and Computation in Human Layer 2/3 Cortical Neurons," *Science* 367 (2020): 83–87. As additional examples of the heroic and versatile work done on understanding the brain, one may look at these works: William Lotter et al., "A Neural Network Trained for Prediction Mimics Diverse Features of Biological Neurons and Perception," *Nature Machine Intelligence* 2 (2020): 210–19; David Zada et al., "Parp1 Promotes Sleep, Which Enhances DNA Repair in Neurons," *Molecular Cell* 81 (2021): 1–15; Goffredina Spano et al., "Dreaming with Hippocampal Damage," *eLife* 9 (2020): e56211. In addition, see Appendices A and B and many references throughout the book. See also the European and US megaprojects with the goal of better understanding the brain: https://braininitiative.nih.gov; https://www.humanbrainproject.eu/en.

stands a glass vessel, a transparent bowl. Its shape is that of a hexagon. In it is some transparent liquid. At the bottom of the vessel lies a brain. Imagine it's your brain! All kinds of electrodes connected to wires have been pushed into its exterior. They are stuck there from all sides.

And this brain, your brain, clearly sees a picture of itself, like a person standing in front of a mirror. But what the brain sees is itself lying there at the bottom of this hexagonal vessel. Most probably, the picture of your brain inside the glass bowl is fed into your brain through those electrodes. About one and a half kilograms of gray matter, which uses up to a quarter of the oxygen intake of the body, and inside it a huge number of tiny electrical connections that would make it to the moon and back if made into one long thread—this truly wondrous creation of evolution that is your brain now sees itself at the bottom of a transparent bowl. In addition, you see some tubes going into your brain. They are probably there to deliver the necessary oxygen and nutrients via some artificial blood, and to take away waste.

For a fruitful simulation to take place, we have to really believe in it, so let us add some rational story to the scene. It's possible that your brain in the laboratory is a last remnant of some nuclear doomsday, or an ecological extinction, a last surviving exemplar that was found by some unknown researchers, perhaps aliens from another civilization, or maybe these researchers simply abducted you and your brain from the face of the Earth. Whatever the case may be, you, living in your brain, in that vessel, clearly do not know the details of how you got there, but you do feel that these researchers are out to satisfy their curiosity regarding our species. In this sense you feel close to these alien researchers, for they are also you; they represent the objective you, the you without a bias-projecting ego, the you who really wants to know who you are, and whether you are alive. You now vaguely begin to see these researchers in the laboratory, moving purposefully in their laboratory around your brain, operating all kinds of instruments in the peripheral areas of what you can see.

Eve, this is what went through my head as I was thinking of how to best approach your question "Are they alive?" which begged the question, "What do we mean by *alive*?" and consequently, "Are we alive?" I found that I had to design some thought experiment that would enable me to examine your question objectively. Let us now be silent for a moment or two and try to observe our brain held captive in that glass bowl. Confucius said that silence is a true friend who never betrays, and for the sake of this simulation, I have a feeling he is right.

We have made the first step in our thought experiment. The more I delved into this thought experiment, the more I equated the transparent bowl in the middle of the alien lab to Einstein's transparent elevators,* with the humongous difference that now the transparency is in the space of consciousness, the subjective space of the mind, a treacherous space indeed.

But immersing ourselves in that hexagonal vessel, full of some transparent liquid, is only the first step, a small opening of the door. In the next chapter we further push this heavy door, which seems to be intent on resisting us.

To conclude this brief exposition of the thought experiment, one may state that a good thought experiment is a get-out-of-jail-free card, as it forces the mind to break through common sense, which frequently is nonsense that we have been accumulating over the years of our life. In other words, a good thought experiment allows us to wander and wonder outside our comfort zone.

---

*I repeat here my previous footnote that I am, in no way whatsoever, trying to even hint at the possibility that the theory of machine death presented in this book comes anywhere close to the absolute genius of Einstein and his theories. Einstein is presented here only as a role model I can learn from.

Part IV

A NEW LIFE REQUIRES A NEW DEATH

5

Going Down the Rabbit Hole

1. THE QUESTION

Dear Eve, in Part I of the book, I walked on relatively solid ground as I was describing an objective observable, predictability, and the experiment measuring it, as well as the technology that enables this profound experiment. I then described in Part II the belief in the *soul* and *free will*, beliefs that go against the notion of predictability, as they are considered by many to be free (unpredictable) agents. I found these arguments to be inconclusive at best, yet it is impossible to completely refute them. The experiment measuring predictability is genuinely profound because it may bring the debate over free will and the soul as free agents to a dramatic end. This is so because predictability is incompatible with the existence of free agents acting within us. But we are still far from reaching the core of the extremely intricate question of human life.

As I was thinking about how to address your question, I first had to identify and list my limitations and biases, which could certainly give rise to me misleading myself. I also had to think of possible techniques to try and overcome these shortcomings of mine.

Specifically, in Part III, I briefly described the preparation I forced upon myself in order to free my mind as much as possible so that I could have some reasonable confidence in my ability to think objectively concerning the horribly subjective topic of human mental life. To our aid, I transported us into the space of a thought experiment, where we may think more clearly. In doing so, we are now nothing but nude brains at the bottom of a glass vessel. We may now start descending, hand in hand, down this dark and slippery rabbit hole. I hope we have equipped ourselves well enough.

We begin to tread carefully as we move down the rabbit hole to suggest interpretations, namely how future results showing a high level of predictability in the human species may be understood. Specifically, I argue that such results warrant a new definition of human life. These interpretations are of course relevant to our discussion on predictability, as they allow us to examine what is really at stake. Indeed, the stakes are so high that, as noted at the end of the previous chapter, every step we now take is a treacherous step.

As we agreed, the question you put forward when you visited my lab, "Are the atoms alive?" requires us first to answer the question of "What is human life?" as this is what you meant when you used the word *alive*, but in fact, conversely, the question "Are the atoms alive?" may also answer the question of human life, for we are our brain, and our brain is nothing but atoms. More specifically, if an atom is just a machine, then it may very well be that we are a machine as well, a complex and convoluted machine to be sure, but a mere machine nonetheless. Our rabbit hole is a circular maze.

Now that we are better aware of the boundary conditions that constrain our thoughts, we can attempt to decipher the elusive enigma of how to best define human life and death, the latter being, as Schopenhauer so eloquently formulated, at the core of everything human.

I go back to the science of predictability in Part V of the book, after concluding this philosophical pause in the coming two chapters.

Let us begin our journey into this maze by clarifying to ourselves: What is actually the question we are addressing when we talk of life? This is the end game of our thought experiment.

Our technological gadgetry is evolving at an ever-growing speed. Each person may choose his or her favorite human achievement, whether it is the smartphone, AI, or reaching Mars. My favorite achievement is already rather old. It is the *Voyager* spacecraft, the man-made object that went the farthest, to the very edges of our solar system, to the edges of the heliosphere, where the solar wind subsides as it meets atoms from the interstellar space. A committee headed by Carl Sagan put on the spaceship some pictures and sounds of us in case someone finds it. I recall I once thought that it is truly unbelievable that so few people love the *Voyager* and so few follow its journey.

With such an amazing rate of progress, it is only natural to ask what new medical constructions, that is, machines and procedures, will enable us in fifty or one hundred years. Yes, longevity will skyrocket, and for those interested, the change of our external bodily appearance will also become mundane. Changing our height, the size of our behind, or the length of our fingers will be done with the click of a button. But could something more profound be expected? I think it can. As described in Part I, AI and Big Data will in the near future provide a never-before-seen look into our core hardware and software. As described in Part VI, this will help design pills that will assist the human race in transcending its present unfortunate state. Namely, as medical instruments become extremely proficient in fixing our biological body, they will then start enabling the improvement of our mental body as well.

But let us once again take a step back. Philosophers have always wondered about the definition of human life. Is a stone that has chemical processes alive? Is a plant, which knows to turn to the sun and, as recently discovered, has electrical currents that resemble a nervous system, alive? Is mold, which contains living material, namely, biological molecules that break the symmetry with their chirality, alive? Is a worm, which has the ability to reproduce, to duplicate herself—namely, to construct something that is so complex yet so organized, in contrast to nature's tendency for maximum entropy, meaning lack of order—alive? Is a human, whose ultimate goal is satisfaction, namely, achieving some chemical balance in the brain, alive? In more popular terms, if a zombie is dead and we are alive, what makes the difference?

We are interested here in a definition that is unique for the life of a human, so universal definitions of *life*, such as something NASA would be looking for to be used in space exploration, for example, *Life is self-reproduction with variations,* or, *Life is a self-sustained chemical system capable of undergoing Darwinian evolution,* will of course not do.

A common answer to this enigma relies on consciousness or self-awareness. We would say that a cow or an octopus should be declared as having the highest form of life, the human standard, if it was found that they are self-aware or conscious. But what is the definition of this self-awareness? What are the criteria for consciousness? Could future medical instruments differentiate between different levels of self-awareness and consciousness even within a group of healthy individuals of the human species?

I would like to give a specific example of what I mean, hopefully without too much loss of generality. It's a letter sent from the large chemical company IG Farben to the commanders of Auschwitz: "We are planning experiments with a new drug. . . . We would be grateful if you transfer to us some women. . . . 200 Mark per woman is an exaggerated price. . . . We suggest not to pay more than 170 per head. . . . We have previously received 150 women. . . . Although very thin, they were satisfactory. . . . The experiments have been done and they all died. . . . We will contact you soon for a new shipment." The guy sitting behind the typewriter must have been well cultured, perhaps even a good piano or violin player. He was definitely a thinker. Was he alive? Those hospital monitors checking the activity of the heart and brain would have surely said he was. Was he self-aware? Was he conscious? If he was not a zombie, in what sense was he not?

"I think, therefore I am" is the famous declaration of René Descartes that comes to mind, but what is the criterion for thinking? After all, robots will soon be able to think as well.

Whether or not we are alive was always up for philosophical debate. So much so that, as I mentioned previously, Schopenhauer stated that fear of death is the beginning of philosophy. But we are not interested here in a philosophical conundrum. In this book we focus on quantifiable aspects of us humans, which instruments of the near future will be able to measure. Specifically, we will avoid words like *self-awareness* or *consciousness* (see Appendix B). These words have contributed tremendously over many centuries to human thought on what human life is, and I imagine they will continue to do so in the realm of the humanities and exact sciences for centuries to come. However, to this day, beyond our subjective feeling that they exist, there is no quantifiable definition of these words, and their vagueness may turn out to be a liability in our attempt to better understand in the near future what the definition of human life is in scientifically accurate, namely measurable, terms.

Confucius said that a human lives twice: the second life begins once we realize we only live once! But perhaps he was wrong, and in the future they will agree that there is also a third life, a life that begins once the human understands that our old common definition for *death* is a charade. A new definition stands at the very core of our thought experiment. It is what awaits us at the end of the rabbit hole.

2. THE PRESENT LIMITS OF OUR KNOWLEDGE

I have already noted that humans have had great success with the science of life and death, and the result is evident in our ever-increasing longevity. In fact, it could be that lifesaving surgery dates back five thousand years. For the sake of appreciating the scientific medical success story in its combating of death, but also its present limits, let us quickly sail across two stories from the early days of methodological medical deduction. These stories and the limits of our knowledge of death which they clarify will help us in our thought experiment to see how AI and Big Data will enable a completely novel and much deeper holistic view of us humans.

The two stories I will now briefly present appear in quite a few essays about the philosophy of science, such as those by Carl Hempel. Let's first recall the work of Semmelweis on sepsis. Ignaz Semmelweis, of Hungarian descent, did the work during 1844–1848 in the Vienna central hospital while he was a doctor in the first maternity ward. He was very distraught due to the fact that many of the women coming to give birth got sick with a disease that was at times fatal. In 1844, 8.2 percent of the women died of it, and in the following two years, 6.8 and 11.4 percent. In the adjacent maternity ward, where nurses did the bulk of the work, the numbers were much lower and stood at 2.3, 2.0, and 2.7 percent, respectively.

He eventually solved the mystery. Although there was no theory to back him up, as they knew nothing about germs, he identified the statistical connection or correlation with the fact that the doctors in his ward also conducted autopsies. He ordered everyone to wash their hands, and the mortality rate dropped to the same rate that was common in the other ward.

It is interesting to note that it was so hard for the medical community to believe his ideas about some horrible material that moves from dead bodies to pregnant women through the hands of doctors that he was ridiculed and eventually died in a mental hospital after having a nervous breakdown. This resistance by the community is nothing but a friction force that appears in the brain every time a new truth overthrows an old one. To minimize the friction, one has to really absorb the idea that truth is infinite, and as we said, the only way to make tiny steps toward this infinity is to constantly attack the prevailing truth. This road is hard for the human psyche, as it sometimes passes right through the center of the horribly smelly market of diverging opinions. But this is a side note, and I brought this story of the work of Semmelweis as a nice example of how logical medical deduction defeated death.

Another beautiful example of the early days of methodological medical examination is the story of a small ethnic group called Fore in Papua New Guinea. The first contact with the West was through missionaries in the 1930s, but a real examination of their culture was only performed in the 1950s. The Fore had a serious disease called kuru, which means a strong shiver. The sick gradually lose coordination in their muscles and eventually die within six to nine months. The first study by scientists was done in 1957. The disease was nothing like science had ever seen. The transmittance of the disease was eventually traced back to the culture of cannibalism common among the Fore, and especially the custom of eating the brain of the deceased. Again, logical medical deduction defeated death.

Both of the above stories deal with biological death. The heart and the brain stop functioning at the very basic biological level, an occurrence represented today by the famous flat lines on monitor screens. Science has also made a significant effort to understand what happens to the body immediately after death. The decay of the body is indeed a fascinating scientific tale.[1] Specifically, if we are able to stop the decay and even reverse it, namely, if we are able to one day bring back the memories and consciousness of a brain that was dead, we will have to completely redefine the border between life and death, even at the very basic biological level. Several recent scientific advances hint toward such a possibility in the far future.[2]

Before we continue in our strides to redefine human life and death, it is important to acknowledge that the biological death described above is as far as science can go, at this point in time, in defining the border between human life and death. Specifically, the definition of this border goes hand in hand with the available medical instruments. It is thus reasonable to assume that as the instruments become capable of seeing far more than they do today, an evolution in the definition of the border between life and death will also be warranted, and indeed unavoidable. As described in Part I, such novel instruments are already here, namely AI and Big Data, and we have simply so far not identified them as such.

—⁂—

Obviously, with the above stories of the medical history of understanding human death, I am only telling half the story about the border between life and death, because science has also made great efforts toward understanding human life, and by that I mean the quantification of consciousness, which stands at the base of what we regard as human life, as opposed, for example, to animal life, or in the words of our Fifth Avenue special lens story (Section 1.2), to zombie life. Namely, scientists have been attempting to find the empirical criteria, which may be measured by instruments such as fMRI, MEG, EEG, and also internal probes,[3] for what constitutes consciousness. In addition to the impressive new equipment, the last thirty years have seen an explosion of new theories, such as integrated information theory (IIT) and the global neuronal workspace (GNW).

However, there are inherent problems originating, for example, from both the phenomenological (examining from a subjective point of view of personal experiences) and epistemological (the theory of knowledge and the distinction between justified belief and opinion) points of view in utilizing consciousness as a criterion for human life. The scientific literature is full of vivid illustrations of how diverging the field is. It is far from any consensus, similar to the state of the fields of quantum gravity, dark energy, and dark mass in physics. It is enough to take a quick look at the Wikipedia entry on consciousness to find sentences like "Experimental research on consciousness presents special difficulties, due to the lack of a universally accepted operational definition," and "For many decades, consciousness as a research topic was avoided by the majority of mainstream scientists, because of a general feeling that a phenomenon defined in subjective terms could not properly be studied using objective experimental methods." Indeed, if you meet a candid cognitive scientist, you will probably hear something like, "It's still just fundamentally mysterious how consciousness happens." (See Appendix B for more details.)

The "hard problem of consciousness" is a term that is generally used by cognitive scientists to describe how difficult the problem of understanding consciousness is.⁴ In my view, the first hard problem in identifying and quantifying consciousness is that it inhabits the same physical space and the same elementary physical processes (e.g., neurons, neurotransmitters) as unconsciousness. So the difference is clearly about the processing, which is a hard problem. Freud used the iceberg analogy to explain that consciousness is just a small part (tip of the iceberg) of what there is inside the skull. Perhaps one of the biggest clues at our disposal is that a big chunk of the brain, the cerebellum, did not develop any consciousness. Another clue may be found in the chemicals possessing anesthetic capabilities that can turn consciousness off. The second hard problem, in my view, has to do with the fact that we are not trying to examine an objective reality but rather a subjective reality. This problem was eloquently put forward by Ludwig Wittgenstein and his thought experiment, which he called "the beetle in the box." Wittgenstein invites readers to imagine a community in which the individuals each have a box containing a "beetle." No one can look into anyone else's box, and everyone says he knows what a beetle is only by looking at *his* beetle. If the "beetle" had a use in the language of these people, it could not be as the name of something—because it is entirely possible that each person had something completely different in their box, or even that the thing in the box constantly changed, or that each box was in fact empty.

Trying to use terms other than *consciousness*, such as *sentience, qualia*, or *self-awareness*, has not improved the situation.

But even in the unlikely event that scientific work will bring the entire community of scholars to agree on what consciousness is and how to quantify it, would there be agreement on utilizing this dreadfully complex mechanistic characterization of the inner workings of the brain as a criterion for the most sacred of all things, human life? Even if one day, far into the future, the scientific community will agree on how to quantify consciousness, this definition would most probably be so complex that it would be beyond what society can adopt as a simple criterion they can try and follow in what concerns their most precious asset, life. While it is clear that the studies of consciousness are extremely useful and important, is there an alternative that is capable of giving us some reasonable insightful answers within our lifetime? As I argue throughout this book, the answer is a resounding yes.

The time has now come to try and look forward beyond the present state of the art.

6

Predictability and a New Definition for Human Life and Death

Eve, we are now deep in this dark rabbit hole, but our eyes are getting used to the dark. We are beginning to see. In the beginning of Chapter 1, we have already mentioned the concept of machine death, and P = 0.5 as an optimal value between a completely random mind and a completely predictable mind (a mathematical formula was also provided). In this chapter we take a long cautious step back and revisit these concepts in a slow-going as well as more elaborate manner and arrive at a verbal definition of how human life could be maximized. All this will also eventually help us answer the question of whether there is a pill that could cure us.

A good chef goes to the market every morning to carefully choose the products that he or she will use to prepare dinner, as they know very well that a gourmet meal is only as good as the products that were put into it. In fact, this is not very different from the well-known wisdom that the output of a computer is only as good as its input. Similarly, the quality of any logical deduction concerning reality, in this case the reality of our brain, is dependent on the quality of the fundamental facts we have at our disposal. Unfortunately, currently we are still missing considerable input concerning the human brain. But as we are looking to the future, we may assume that the instruments of the future will have plenty of good input concerning internal brain processes, with the next generations of machines going far beyond the present capabilities of machines such as MRI and EEG. We may also expect to have good input concerning observed brain outcomes by measuring and quantifying human activity, for example, through our digital footprint.

However, input is not everything. Just as a gourmet meal is also dependent on cooking skills, and the output of a computer is also dependent on the quality of the software, deciphering the reality of human life based on the above input concerning humans is only as good as the fundamental definitions we construct to categorize and interpret that input. After all, we humans constantly decide what reality is by comparing the input we get through our senses to our internal classifications of what each input means; thus, these classifications or definitions are of paramount importance, as they stand at the base of all that we understand. So now, for the first time in our thought experiment, let me formulate such a new definition of human life and death. We can then use this new definition, together with available data, to come to some conclusions as to who we currently are and where we can, or want to, go.

So, what can such a new definition of the border between human life and death be? Let me first look at life, the opposite of death. I start with Jean-Paul Sartre's famous statement that a person is just a projection of thoughts into actions, and there is no reality other than that formed by acts, so a person exists only when he realizes his thoughts in action. Consequently, at the end of the day, a person is nothing but the sum of his actions.

If action is the only test of a person, can we consequently assume that actions may also be the criteria of human life? Namely, that there would be some set of well-defined actions beyond reproduction and other actions of mere survival, such as helping an old lady cross the street, that if carried out by an individual would indicate that he or she is alive? Many people would not accept this to be the case, as they will argue that even a robot can be programmed to have an exquisite character giving rise to adorable actions, where by robot they are referring to an entity that is devoid of human life.

Indeed, in these modern times of ours, many intuitively view robots as entities best representing that which is devoid of human life. Thus, the word *robot** is utilized to portray the opposite of life. It seems robots (or androids or

---

*The word was invented in Prague by Karel and Josef Čapek in the 1920s, and may have originated from the ancient Jewish legend, well known in Prague, of the Golem, a humanoid typically made of clay or mud that was given life and then destroyed when it became too dangerous (indeed, I found a researcher claiming that the brothers first considered using the word Golem for their human-like machine). On a personal note, I allow myself to add a rather bizarre anecdote, which is that there is

humanoids) are the popular choice for contrasting with human life as they are especially akin to us humans. Consequently, many feel that the definition of human life should be aimed at differentiating between us humans and robots. As I argue in the following, this strategy of defining human life by differentiating us from robots is likely to be a futile avenue if our goal is to define human life in a meaningful way, beyond merely having metabolism or an evolving (mutating) DNA, or a biologically based functioning heart and brain.

The difficulty presented to us by robots when we come to define human life has been dwelled upon by many, including in cult movies such as *Blade Runner* or more recently TV series such as *Westworld* and *Black Mirror*.* The idea that our brain is simply a computing machine had already started to percolate a long time ago with theories like the computational theory of mind (CTM). If our brain indeed just computes, or in the words of Section 2.2, is simply a complex feedback loop, then the road is open for robots to be us. Whether or not these robots will have in their heads digital silicon chips (regarding which John Searle, with his Chinese room argument, claims they cannot have consciousness, or even intelligence, even if they do pass the Turing test†[1]) or something more advanced[2] is of no consequence to us here. I believe there is no reason not to assume that the time will come when artificial thinking machines can reliably emulate the brain via a good-enough analogue, or perhaps even reliably simulate the exact mechanism of our brain, and thus have the same consciousness that we have.[3]

To help us visualize the fundamental difficulty in identifying us by contrasting us to robots, let's recall the story of Data, an android serving as an

---

significant evidence that the first printed version of the Golem story was printed in 1909 in the Hanoch Folman printing house (the grandfather of my father) in Piotrokow Trybunalski near Warsaw, Poland.

*One should not confuse the issue we are discussing, of whether robots will ever achieve the status which we define as human life, with the other, more popular problem posed by robots to our very existence. See, for example, *Human Compatible* by Stuart Russell (2019) or *Superintelligence* by Nick Bostrom (2014).

†Roger Penrose presents similar thoughts in *The Emperor's New Mind*. I myself am not sure if a collection of well-defined elements (i.e., transistors), executing a digital program, can accurately emulate the brain with its analog features in time (no clear computer clock) and space (the brain is rather vague concerning which neurons contribute to which process) and with its connectivity loops, which make it unclear which neuron is affecting others and which neuron is being affected by others (a neuron may even affect its own input), so the whole process of cause and effect becomes extremely complex. See also the unfolding argument against causal structure theories in Appendix B. Finally, see the warning about the future capabilities of AI from one of the founders of the field, Nobel laureate Geoffrey Hinton, in the *MIT Technology Review* (2023): https://www.technologyreview.com/2023/05/02/1072528/geoffrey-hinton-google-why-scared-ai/

officer on the flight deck of the Starship *Enterprise* in the TV series *Star Trek*.* He looks like a human and talks like a human. This specific android was extremely advanced and unique. His brain had memory and computing powers never seen before. In fact, in all intelligence tests, he ranked much higher than humans. Aside from this specific android, no others were made, as his inventor died and the designs were never found.

When Natasha, the security officer on the ship, died, the android felt feelings of longing and deep sorrow, sensations his inventor probably did not even imagine could be aroused in his creation.

One day a special emissary arrives on the ship with an order from the high command: The android should report to the fleet headquarters so that he can be dismantled and studied, with the aim of building many more like him. The chances that this specific android would ever function again were slim, as the knowledge regarding his delicate mechanisms were scarce at best. Legally speaking, the android was the property of Fleet Command, and they had every right to do as they pleased.

The android asked his direct commander, a human, to appeal. First, argued the android, he had crossed the line between a machine and a being with self-awareness and feelings, and therefore he should not be denied the basic right given to every human to live. Otherwise, he emphasized, he would be turned into a man who is below other men, namely a slave, and slavery is illegal by law. Secondly, he argued, such a move would be the doom of the human species, since the creation of a race of servants and slaves, even if the latter are deemed a machine, will affect the life of the human as if the slave were flesh and blood. The human is no longer required to be brave, to sacrifice, to overcome challenges, and perhaps even to possess human solidarity, since the origin of the feeling of solidarity is in the collective need for safety, assistance, and warmth. The story goes on, but let me stop here so we can rush back to our real-life story, while each of us may invent our own ending to this wonderful *Star Trek* episode.

If the exquisite actions of the robot cannot differentiate him from a living human, perhaps his thoughts can? However, as robots will be able to think, the famous declaration of René Descartes, "I think, therefore I am," is not very helpful. The definition of life by what we call self-awareness or consciousness also falls short of separating humans from robots, for, as noted, AI

---

*If you read journals like *Nature Machine Intelligence*, you become aware of how quickly we are progressing toward making this science-fiction scenario a reality.

is indeed expected to acquire these features, if not with digital algorithms then with accurate emulations of the human brain.

Later, by the way, Descartes added, "I doubt, therefore I am," as for him one cannot be said to be truly thinking without doubting. But even skepticism, which as we discussed is the foundation of every free mind able to investigate, is not a good differentiator. A robot may be programmed to imitate skepticism, and a higher level of AI may even acknowledge that it does not know everything and consequently must follow a skeptical approach to its own deduction.

A religious person may even state that the belief in God may differentiate between the all-rational AI in the head of a robot and a living human. We may therefore ask ourselves if a computer can believe in God, even when it has no proof of his existence. I think it could. If it's a primitive computer, it will believe if it is programmed to believe. If it is a computer with a personality, it will believe if it wants to imitate people so that it becomes one of them, perhaps out of envy of, or respect for, its creators. It will certainly recognize that in order to do so it must also adopt irrational human traits such as believing in God. Out of respect, it will most probably not even inquire as to the definition, or lack thereof, that humans have for this God. If it's a really advanced computer so that it weighs its unknowns, it will certainly believe in God, as it would adopt Blaise Pascal's wager, which states that it makes much more sense to believe in a god, since if you believe and he doesn't exist, you have lost very little, and if you believe and he does exist, you win eternity in heaven. On the other hand, if you don't believe and he does exist, your losses are immeasurable.*

Finally, if one is intent on differentiating us from robots, one may simply say that we will never be robots and robots will never be us, as they don't have biology and chemistry, they don't evolve and mutate, they don't have metabolism, and so on. I believe these arguments to be superficial and shortsighted; eventually humans will take and insert into their bodies and brains what is best in robots (a human enhancement trend called cyborgs), and similarly, in robots we will put what is best in humans, even a partially biological brain if we come to the conclusion that it has some advantages (e.g., what we call intuitive thinking). Some call this coming fusion robo-sapiens.[4] With the advent of advanced bioengineering, it could even be that we will eventually

---

*I personally think that the working assumption "If you believe and he doesn't exist, you have lost very little" is fundamentally wrong. As Pascal was a devout believer, I can understand why he would make such an error.

make some robots identical to humans in their internal biology (e.g., for survival in hard electromagnetic environments, or simply for companionship). As human technology will always aspire to gain the best of all worlds, I believe it stands to reason that the two species will one day be one.[5]

For our purpose here, one may thus assume that eventually there will be no observable (measurable) difference between robots and humans, so trying to define human life in a way that differentiates us from robots, or androids, or humanoids, namely, using these names in an effort to define human life by contrast, is futile. In fact, it may very well be that whatever new definition we find for human life will eventually be used to define robot life as well, as they move away from being rudimentary machines.

Nevertheless, before continuing the search for new words that could be used in a novel definition for human life, I would like to note that discussing robots as I briefly did above is helpful, as they seem to be a great mirror that will become better and better at showing us our own image, a sort of magic reflector forcing us to ask questions and scrutinize the very foundation of our species. In other words, when defining human life, using robots is a highly insightful philosophical exercise, helping us to distill who we are, and, more importantly, who we want to be, and specifically what we want our life to be. Perhaps this will eventually be the greatest contribution of robots to the human species.

Eve, my dear, it seems we find ourselves having made no progress, as the question remains: How do we define human life? As noted, the standard definitions of today are extremely flawed, as they are based on words and concepts such as *consciousness*, which have never been defined precisely enough that they could be quantified and measured in an agreed-upon manner. As the instruments of the future will be able to probe us and collect data on us with an accuracy and magnitude that currently we cannot even begin to fathom, our definition of human life will have to evolve as well, since today's definitions will quickly become archaic.

Realizing how very hard it is to define human life, I was contemplating that perhaps I could still arrive at a definition from the back door, by agreeing on what human life is not, even if I cannot use the notion of a robot to define the opposite of human life.

The next step was therefore to try and define human death. In searching for common ground, we are after a definition that is likely to be accepted by all, irrespective of education, age, religious beliefs, race, socioeconomic status, or gender. As language is the foundation of our collective thinking and thus may also constitute a great barrier, we must do our best to choose the best possible word or words. Death is associated, on the one hand, with morbid words such as *decay*, *end*, and *nothingness*, and on the other with hopeful words such as *afterlife*, *soul*, and *reincarnation*. Unfortunately, none of these words may help us with an operational definition. I thus told myself that I must continue the search for another word that could represent death in some operational, measurable manner, irrespective of whether the heart and brain are biologically functioning or not.

"What about the word *machine*?" I one day asked you, namely, the avatar Eve roaming around inside my skull. Perhaps we could all agree that life is the opposite of machine, because when we say *machine*, we also mean an entity that is not alive. But why exactly does our intuition attribute this meaning to the machine? What is the definition of a machine that makes us smell death?

It seems the most fundamental definition of a machine lies in its predictability. In fact, if by machine we are referring to classical, nonintelligent machines, the better the machine is, the more predictable it is. A good machine is predictable. Namely, if it works as it was designed, we should be able to know its exact state for any specified moment of time in the future. For example, as I noted in the introduction, as long as the amounts of air, water, lubricant, and gasoline going into a piston engine are well defined, the position of the piston in its cylinder housing at any time in the future is well known. It is this predictability that enabled humans to get to the moon and Mars, and to build smartphones and computers.

So, if human life is the opposite of machine, and machine is predictability, can the definition of human life simply be an entity that is unpredictable? This cannot be the case; the outcome of a roulette wheel or a throw of the dice is random and therefore tantamount to unpredictable, but I think we would all agree that our definition of human life cannot be equated to a throw of the dice. This is not only wishful thinking. The significant stability of opinions exhibited by human brains, from what color or clothing the brain likes to what politics it prefers, clearly shows that something quite different from random decision making is taking place. In addition, as we are aiming here at a definition that also represents what we would like to aspire to be, we may discard the random, namely, the completely unpredictable extreme, as unwanted. For what is really left of our species if we are nothing but a throw of dice? Thus, when it comes to a new definition of human life, randomness, like predictability, is also the opposite of life; namely, it is death.

It seems we have little option but to search in the center between these two extremes of predictable and unpredictable. Our definition for a person who is alive can thus be *unpredictable but not random*. Your avatar then whispered to herself, "UBNR ... UBNR ...," as if she were letting some interesting wine spiral around her tongue.

Namely, I then argued to her, to differentiate ourselves from a rudimentary machine, we would like to be as unpredictable as possible, but we would like to do so without turning into completely random beings. Consequently, if our instruments of the future are able to test, by monitoring a person's actions, whether or not a person is in the center between unpredictable (random dice) and predictable (machine), namely, how close they are to $P = 0.5$ (as described in Chapter 1), then these medical instruments of the future will be able to tell us if the person is mentally alive, or simply alive according to our more fundamental definition of human life.

Let us make a small cautionary statement by going back to the piston-engine example discussed above. The Eve in my head responded by contemplating: One may say that a badly engineered engine also lies in the middle between random and predictable, but a bad engine is clearly not alive. I answered that, indeed, such a counterexample forces us to declare our new criterion for life to be, in the language of logical arguments, a necessary condition but not a sufficient one. But even as a necessary condition, our definition for human life serves its purpose. She nodded.

Measuring random behavior is not that hard (e.g., through testing for temporal correlations), and in fact that's how present-day machines evaluate the quality of random number generators (used, for example, in the gaming industry or in quantum communications). This means that you cannot use any knowledge from the past to predict the future, as the two are completely uncorrelated. In a random number generator, the number that just appeared a second ago does not give you any advantage in trying to figure out what the next number will be. If you find that there are no correlations, you know it's a random phenomenon.* But in any case, as the rate with which we change our opinions, from politics, to art, to music, is quite low, it seems we live much closer to the predictable edge than to the random unpredictable edge, so in this book we focus our attention on predictability.

But can predictability be measured? As we have shown in Part I, and as we detail in Appendix C, the measurement of predictability has already begun. A reasonable extrapolation of existing AI and Big Data technology tells us that the immense power of AI and the all-encompassing data harvesting will indeed, already in the near future, be able to measure our predictability in a technologically mature manner. In Part I we defined a parameter, P, with a value between zero and one, to represent predictability.

This is the epicenter. This is ground zero of our extinction as a race of free entities. If we accept that predictability is death and that predictability may be measured, and, furthermore, if our species is starting to show high levels of predictability, then we are doomed.

To summarize, our definitions for human life and death thus replace ambiguous terms such as *self-awareness* and *consciousness* with the clarity and decisiveness of a measurable observable, predictability P. Being in the middle between the two extremes, namely, P ~ 0.5, means *unpredictable but not random*.

Can we nevertheless find some connection between our new observable, predictability, and the traditional term used, *consciousness*? I think we can, and I have briefly explored this question in the introduction, where I noted the example of mind-altering drugs. On the one hand, cognitive scientists tend to agree that hallucinogens do take you to a higher level of consciousness, and on the other hand, many would also agree that mind-altering

---

*In mathematics, the analogue of a random number generator is a choice sequence, or a lawless sequence. See the work of L. E. J. Brouwer.

drugs induce divergent thought, namely, creativity, which is tantamount to less predictability. (It is not by chance that so many artists find it helpful to use hallucinogens; see also the section on creativity, Section 9.3). So, the two seem to be connected.

We have thus found a simple parameter, which may be feasibly measured by instruments in the not-so-far future, and which has a deep grasp of the amount of human life in an individual. Such an observable relates to the actions of a human being, namely, to the output of the brain.* It is thus much more reliable than trying to identify life through elusive notions such as consciousness, measured, at least partially, by the yet even deeper vagueness of complex internal brain signals.

We may now further examine this new definition of human life and its consequences, as well as the most important question: Is there a pill that could heal us? I do so in Part VI of the book. But first in Part V, in Chapter 7, I go back to science and address possible ways in which nature may be fighting determinism (i.e., predictability) in our brain. Next, in the beginning of Chapter 8, I introduce a simplified two-layer model of the brain, which helps us visualize the internal battle between an unpredictable brain and a predictable one. The above understanding will be most helpful when we come to engineer the pill. Finally, I conclude Part V with a summary of all that has been discussed thus far.

---

*The fact that we rely here only on the actions of a person when we come to probe him or her, rather than trying to measure internal brain signals, is reminiscent of the philosophical theory of consequentialism. But whereas the latter is concerned with the moral consequences of an action, here we are concerned with the consequence of the value of P, namely, with whether an external entity, specifically AI, could have predicted the act.

Part V

THE FUTURE OF PREDICTABILITY

7

Can We Undo Predictability?

I believe, Eve, that we should now end the philosophical pause we took, in which we attempted to interpret the near-future experimental results in the measurement of predictability in terms of a new definition for life and death, and journey back to the science of predictability. Even if one finds it hard to accept the proposed definition for life and death, the science of predictability still stands. One may choose his or her favorite interpretation of the experimental results, but if indeed, as I conjecture, a high level of predictability will be found in humans in the near future, the whole of humanity will have no choice but to consider the profound implications and consequences.

Specifically, a high level of predictability will put to a dramatic rest numerous past arguments in favor of freedom, such as the idea of a *soul* or the notion of *free will*, presented in Chapter 2. Examining these arguments that suggest the existence of some agent of freedom within us, we found them to be weak or at the very least inconclusive. The question may now arise of whether more modern and conclusive arguments could be put forward suggesting that there might still be mechanisms capable of fighting predictability. This is the first step in concocting some cookbook recipe for a pill that could perhaps cure us.

1. CAN INTERNAL NOISE SAVE US?

Can internal brain noise save us? Namely, can internal noise undo predictability? Yes it can! At least partially, at least for some people, but something is better than nothing. Let me explain how.

135

Putting aside tall orders such as indeterminism due to the soul, or freedom to choose (i.e., free will), or even something more down to earth such as unpredictability due to emergent phenomena, and observing the basic functions of our brain, we find that an extremely mundane phenomenon such as noise may be a game changer in the story of the brain as a predictable machine, as when we say noise, we mean unpredictable (random, stochastic) fluctuations (in the following, *noise* and *fluctuations* are synonymous). When physicists say *noise*, they simply mean a phenomenon that cannot be described and thus predicted by some simple formula, irrespective of the origin or representation of this phenomenon. In the brain, these fluctuations are electrical fluctuations, spreading random electrical noise throughout the neural network.

So in fact while noise does not deliver the sought-after free will, or the metaphysical soul, it may just deliver enough potent randomness to save us from the death of being completely predictable by some AI in the near future (treating our brain as a black box, the interior of which is unknown to AI; see Chapter 1) or by some supercomputer in the far future (which actually follows the internal processes of our brain), at least until such time that the supercomputer becomes extremely powerful so that it can also calculate and predict these fluctuations.[1]

Let me briefly explain that such fluctuations are not truly random, as they are still part of the deterministic and predictable non-quantum world—sometimes referred to by physicists as the classical world (see next section on true randomness inherent to quantum theory). Nevertheless, to predict these fluctuations would require immense computing and memory power, not to mention monumental knowledge of the initial conditions of the system (as well as strict isolation of the system to avoid external disturbance). You would really need to be able to follow every atom in the brain (see Section 1.3 on the human machine). Hence, we may say that for all practical purposes (FAPP), these fluctuations may be thought of as being random and unpredictable by foreseeable technology.* Consequently, noise has the ability to pull us away from the machine death indicated by $P = 1$.

---

*One may of course make it even harder for the supercomputer by assuming that external fluctuations also impact the brain, such as fluctuations in the light coming from stars, but it stands to reason that these are weak perturbations on the brain.

—⚉—

No one really knows what the source of the observed noise in the brain is. It is fascinating to reflect on whether this noise simply comes from some environmental causes, such as a thermal origin as explained below, or other environmental causes, such as changes in blood flow and its chemistry (e.g., how much nutrients and oxygen are being delivered to the brain), and evolution had to adapt, namely, to develop a brain that can work properly with such fluctuations. Alternatively, it could be that evolution made a brain that predominantly produces noise inherently for some advantage. What is clear is that there is noise and, if we trust evolution, that it is in some way beneficial.[2]

Let me note in passing that, interestingly, one clear advantage of noise, as was emphasized earlier, relates to our ability to learn. As noted, studies of decision-making processes talk about exploit versus explore. Typically, when we know how to get a reward, we act on it, namely, we exploit the situation, but sometimes scientists find that even if the environment (stimuli) is extremely clear and simple, decision-making processes can take a wrong turn. They call this lapses in judgment. Some speculate that this is not a bug in the system but rather a smart way to further explore the situation in order to better understand it, and perhaps later on get even bigger rewards. These occasional lapses in judgment may be driven by noise. Obviously, for some local electrical neural noise to get to the level of decision making by the whole brain, it must be amplified (as opposed to being suppressed by averaging with noise from other local areas). Indeed, at this macro level, scientists are working hard to identify with neuroimaging which processes mediate the exploit-explore trade-off,[3] but here we will not venture into these complex mechanisms, which seem to even depend on age.[4]

Importantly, I have noted that some scientists, although not having definite proof but after decades of experiments, are willing to state that they are convinced that some form of freedom exists and may even be enhanced through our own actions. For example, Kornhuber and Deecke—who discovered in 1964–1965 the *Bereitschaftspotential* (BP), or readiness potential (RP), a brain potential in the EEG that precedes all our willed movements and actions—state that a person has "relative freedom." Could this "relative freedom" be related to the ability to explore, and following the above-discussed connection between exploration and noise, could this point to a bond between freedom and noise?

The idea that noise is connected to the survival of a somewhat nondeterministic (i.e., not predictable) or perhaps to some extent even free entity inside of us has been starting to percolate. Specifically, noise and its fluctuations are thought to give rise to internal ignitions of thought processes. Some cognitive scientists conjecture that these fluctuations are so important that they are responsible for what they call free behavior, which they associate with brain outcomes such as creativity.[5] They define free behavior as "the group of behaviors that are not fully determined by an external stimulus or instruction." Namely, this behavior is not (fully) a part of the automated, even if complex, feedback loop we discussed in previous chapters. In other words, this part of our behavior is not entirely a response to something external but is rather initiated from within. I suspect that they are using "free behavior" rather than "free will" because they have no claims that there is real freedom involved, in the sense that not everything is predetermined by the determinism of nature. Nevertheless, we should agree that, as noted in the beginning of this section, such fluctuations may indeed have significant consequences.

So, what proof do the proponents of free behavior as arising from fluctuations bring? In research by Norman and Malach,[6] they focus on measurements done with a device mentioned earlier, the intracranial EEG (iEEG). These are electrodes that are implanted inside the brain. (See also their early work.[7]) In their experiment they ask human subjects to freely recall some images they have seen before in some pictures or movies they were exposed to. Hence this is not purely associative, out-of-context thinking, but not far from it. The findings were that, "throughout the recall period, one can observe that the neurons' activity slowly fluctuates at a low amplitude in an apparently spontaneous manner, until, just before the recall, one of the slow fluctuations becomes large enough to cross a cognitive bound or threshold—allowing the patient [human subject] to become aware of the spontaneously generated mnemonic content, and to verbally report about it." Indeed, these researchers also report the observation of increased noise amplitude when we try to think independently.[8] They conclude that the random (spontaneous) fluctuations "contribute to the emergence of free thoughts and spontaneous actions."

Although, as we noted, correlation is not necessarily causation, so that complete proof is not yet available, it appears that significant support has been found for the important role noise plays in the creation of unpredictable thoughts.

I briefly add that some scientists believe that an interplay between stable and chaotic processes is also extremely important to brain functions.[9] In the context of our discussion, chaos, like noise, may be thought of as being FAPP random and unpredictable by foreseeable technology, so chaos may also deliver enough potent randomness to save us from the death of being completely predictable by some AI in the near future. In fact, noise and chaos may be working together, as chaos may be thought of as an amplifier of noise. While noise slightly changes the initial conditions, chaotic processes take these small fluctuations and enhance them (amplify them) to very different final outcomes.

Eve, in case you are interested in knowing more about the fascinating topic of noise, I now further clarify what I mean by *noise* and how it is measured. If, on the other hand, you feel you understand enough, please do hop to the last four paragraphs of this section, starting with "In summary . . . ."

Scientists obviously like to measure things. If, for example, you measure a current in some electrical circuit and the needle in your very accurate ampere meter jitters during the measurement, then you have some jittering in the current, and we call this noise. If you look at the screen of a laboratory oscilloscope measuring the voltage on a resistor in this circuit, you will see voltage fluctuations in the form of waves, where the height (amplitude) of the waves is the measured voltage and the horizontal axis is just time. Unlike regular ocean waves, these waves on the scope screen are typically quite messy, without any clear periodicity. We thus say that the noise is made up of many wavelengths or frequencies. You can do a mathematical procedure called a Fourier transform to identify exactly which frequencies you have and with what intensity (weight). Different forms of waves have received different names over the years, such as white noise (where all frequencies are equally present) or colored noise (where some frequencies are more present than others), or the most common noise, which is called $1/f$ noise (or flicker noise), where f means frequency (measured in units of hertz or Hz, namely, the number of wave cycles per second) and $1/f$ is the strength of the noise, so that this noise has mainly low-frequency components.

Depending on the system being measured, many different processes can give rise to noise. For example, a famous noise named the Johnson-Nyquist noise appears in electronic circuits even when there is no voltage or current being applied. The electrons in the metal wires simply move in random directions due to the temperature of the metal. As we know, temperature gives rise to velocity if a particle is free to move (e.g., in the atmosphere), and in the metal, some electrons are free to move (and that's why we can push a current through metal wires). The electrons frequently scatter from material inhomogeneities and change their direction. This thermal velocity of electrons and their frequent changes of direction then give rise to fluctuating magnetic fields, another type of noise that we can measure. We should also note that the above thermal energy gives rise to a wide distribution of velocities even for a fixed temperature. The general topic of noise in nature is fascinating, but we will not elaborate on it further here.

Concerning the $1/f$ noise, which is a good example of an omnipresent noise, it has been measured in numerous systems, starting with a vacuum tube in 1925, semiconductors in 1974, and the human heart in 1981. But does such noise, or any noise for that matter, exist in the brain? Indeed, it does. Noise in our brain is a well-documented phenomenon.[10] In the late 1990s, the first experimental hints concerning $1/f$ noise in the brain started coming in from MEG[11] and EEG[12] measurements.

The importance of fluctuations has been emphasized in several studies.[13] I would like to give you a sense of what such a scientific paper looks like, so instead of just leaving it as a reference, I copy here a short excerpt from its introduction. It states (see references within):

> Particularly for sensory systems, it was assumed that in the absence of a stimulus (e.g., in complete darkness), the sensory cortex enters an uninformative, low-level baseline mode. However, following the pioneering research of Arieli et al. (1996) in anesthetized animals, paralleled with BOLD (blood-oxygen-level-dependent) fMRI recordings in the human motor cortex by Biswal et al. (1995), numerous studies have, by now, established that in the absence of a stimulus or task—in what has been termed the "resting state"—cortical networks enter into a highly informative mode of spontaneous activity. In contrast

to the active mode, these so-called resting-state fluctuations are characterized by low-amplitude modulations of activity and ultra-slow (<1 Hz) dynamics. Although most research on resting-state fluctuations has relied on BOLD imaging, subsequent research has confirmed their existence also in firing-rate modulations (Nir et al., 2008), ECoG recordings (He et al., 2008; Nir et al., 2008), and scalp EEG (Schurger et al., 2015), following on the seminal studies of Berger and colleagues (e.g., Niedermeyer and da Silva, 2005). Importantly, the slow spontaneous fluctuations are not confined to specific resting-state networks but emerge in each and every cortical site that has been studied so far. For example, the human visual cortex shows widely spread and highly structured spontaneous BOLD fluctuations in the absence of visual inputs (e.g., Arcaro et al., 2015; Nir et al., 2006). The structure of the spontaneous fluctuations appears to be highly informative and likely reflects the statistics of the natural environment and cognitive traits (Berkes et al., 2011; Harmelech and Malach, 2013). The structure and ubiquity of these fluctuations suggest that they may constitute an important mode of cortical function.

On the fundamental importance of noise to the very core of who we are, one may take a look at the references in the notes.[14]

In summary, it seems that these proponents of fluctuations as the base for free behavior are using the word *free* rather loosely, but I believe that their intention is correct. While nothing is really "free" in a deterministic nature, the thoughts are "free" in the sense that they are not an automated reaction of a feedback loop to input but are rather initiated independent of external stimuli. As there is no input, there is also no action by a feedback loop, and the outcomes are much harder to predict.

But why did I start this section by noting that it could only help some people? The researchers themselves concede that the effect of these fluctuations must be limited by boundaries and constraints determined by personality traits and other predetermined features of the brain; otherwise our behavior would truly seem random, and our feedback loops, so important for our survival, would not be able to function. It stands to reason that genetics, conditioning, and education all somehow determine the eventual potency of these random processes to affect the brain at the decision-making level. I go back to this important discussion when I present a two-layer model of the brain in Chapter 8.

To conclude, noise does not award us with real freedom, as it is part of the deterministic universe. On a philosophical level, in this scenario we are still machines, albeit noisy machines. Noise does not even make us truly unpredictable, but it does have the potential to make us unpredictable FAPP. A great question to be answered in the future is whether we ourselves have the ability to increase its potency, namely, to purposefully increase the level of noise in our brain, thus reducing predictability P. This is the beginning of the pill!

However, when a supercomputer comes along that can faithfully represent the Laplace demon in great detail, even noise will have lost its charm, as the demon should even be able to predict the noise. In any case, the level at which noise currently affects our brain at the decision-making level could again be measured by our omnipotent parameter P.

2. THE QUANTUM MIND

If quantum processes are dominant enough in our brain to affect our thoughts, the randomness of quantum mechanics may introduce true random noise into our deterministic mind so that even some imaginary all-powerful future supercomputer—discussed in the previous section—cannot predict it. This is because quantum theory has an inherent random part built into it (which is why it's called a statistical theory). A specific set of initial conditions may lead to several possible outcomes. This is so at the most fundamental level of nature and will introduce some unpredictability, not just FAPP like classical (non-quantum) noise does (see previous section). Consequently, even the Laplace demon, which we discussed as representing a deterministic nature, would not be able to follow and predict quantum randomness.

For example, if a photon hits a piece of glass that is coated with a thin layer of silver so that it is only partially transparent, quantum theory tells us that there is no way to predict whether the photon will go through or be reflected. The theory can only predict what will happen statistically for a large ensemble of photons. Actually, a new generation of random number generators used by the gaming industry, gambling sites, and secure communication systems are based on this quantum randomness.* The simplest true random number

---

*Different from the quasi-random number generators each of us has inside our computer, which are not truly random but rather are based on a formula, quantum random number generators are truly random, as they are based on quantum randomness.

generator you can build is based exactly on the above single photons impinging on such a partially transparent piece of glass.

This true randomness has dramatic consequences. We have already stated that life must be in the middle between randomness and predictability. If the brain is generally deterministic but quantum mechanics introduces some true random noise, it may indeed put us in the middle, namely in the regime of *unpredictable but not random.*

Dear Eve, the question of whether the brain has some quantumness left at the decision-making level or not is of crucial importance, so I allow myself in the next few pages to go into the physical details as well as some speculations. As is our tradition by now, if you feel disheartened by such detail, I implore you to skip over these details and go directly, without any misgivings, to the fourth paragraph before the next section, Section 7.3, a paragraph in which I also address your original question on whether the atom is alive (the paragraph begins with "Lastly, Eve, . . .").

The question of whether quantum processes may be dominant enough in the brain to affect the brain's outcomes (namely beyond the functioning of individual atoms or molecules) is a hard problem to solve. The first answer is no, since the destructive forces of decoherence (namely, the opposite of coherence, where coherence is quantum dominance) are very strong in the ambient conditions of the brain material. Indeed, as noted previously, some people, for example, Max Tegmark,[15] have calculated that the number of atoms involved in the firing of even a single neuron, and the type of coupling to their immediate environment, is enough for quantum theory not to be relevant. However, a closer look may leave the door for the quantum mind open. This is because when you calculate the strength of the destructive decoherence, named in the professional jargon the *decoherence rate*, you must put into your calculation the parameters of the very specific system you have in mind. Indeed, some claim that in the above-mentioned calculation the numbers and assumptions which were used were wrong.[16] Furthermore, in the calculation, only two specific scenarios were taken into account, that of a normal neural pulse mediated by ions and that of the microtubules suggested by Roger Penrose. Numerous other mechanisms that would allow the quantum mind to operate may be conjectured. For example, some hypothesize that light may play an important part in the brain and that axons may act as optical fibers.[17] Experts in a quantum effect named superradiance, governing the emission of photons from ensembles of atoms, speculate that there may be a connection to coherence in the brain.[18] Other possible systems with lower decoherence rates that may enable quantum information processing in the brain are neutral particles (as opposed to the charged ions typically traversing the axon) and spins, especially nuclear spins (angular momentum of protons and neutrons), which are known to have much lower decoherence rates than electron spin (see discussion about spins in the following).

In addition to a variety of possible quantum information mediators, there are also possible mechanisms in which the environment could be fine-tuned to protect the system against decoherence. In fact, recent findings in the new field of quantum biology seem to claim that evolution has found ways, to which we are still not privy, to safeguard against the hindering decoherence processes. For example, there are persistent claims that the very high efficiency of the photosynthesis process of converting light to usable energy in plants may only be achieved by long-range quantum processes (namely, beyond the scale of a single atom or molecule). There are those who even claim that some quantum coherence is maintained in the retina of birds, and this is how they see Earth's magnetic field and can navigate (most probably through a process called Larmor precession). If this is so, then nature has found ways to protect its delicate quantum processes against decoherence. I guess the only real proof of this will be available when experimentalists show that they can artificially imitate such protection mechanisms in the lab. In any case, if this is true, then there may be similar defense mechanisms in our brain.* (See next section for more information about possible quantum processes in the brain.)

---

*It is interesting to note that although quantum theory has randomness at its core, it may still lead to a deterministic machine. This is proven by the apparatus we call the quantum computer. If it would not be deterministic in its final output, it would not be called a computer. So if the brain has dominant quantum processes, it may either produce random noise, which, as stated above, may save us from strong determinism, or it may still be part of strong determinism. If somehow evolution managed to suppress decoherence on the large scale of the brain, the second option would require that the brain really fine-tune the dynamics of the quantum system so that it is a hybrid system that also includes classical operations like measurement, just like in a quantum computer, in order for its outcome to be deterministic in the sense that the same question always eventually gets the same answer. No such fine-tuning would be needed in the case of a superdeterministic world, à la John Bell (see footnote in Chapter 2, in the section on free will). But again, one needs a mixture of randomness and deterministic predictability to reach $P = 0.5$.

To better understand this issue of decoherence, let me start from the very beginning of what quantum theory is all about. Don't worry; this will be brief.

Perhaps we can start with the story of the ultraviolet catastrophe. At the end of the nineteenth century, scientists discovered they had a big problem. When we heat up a solid, it emits electromagnetic radiation. It emits this radiation, which is also called blackbody radiation, in many colors, that is, in many wavelengths. It's quite amazing that this radiation hardly depends on the material itself. As the object gets hotter, the main radiation is emitted at shorter wavelengths. That's why a hot iron seems red, but an even hotter iron seems yellow. It is also the method used to know the temperature of distant stars. They call the curve describing the amount of radiation at each wavelength the spectrum. The problem was that the spectrum they calculated from known theories was completely different from what they measured in experiments. The error was large mainly at short wavelengths, beyond the color blue, known as ultraviolet, hence the name, the ultraviolet catastrophe.

A smart guy named Max Planck solved this problem in 1900, but he had to make a crazy assumption. He assumed that the atoms and molecules in the material, which vibrate due to the heat, cannot just have any amount of energy but only discrete amounts. That seemed a ridiculous assumption, as we know that objects in our world can have any amount of energy. For example, cars can go at any velocity we want them to go. He called these discrete amounts of energy that the particles could receive "quanta."

The next interesting thing in the birth of quantum mechanics perhaps happened in 1914. Two good experimentalists named James Franck and Gustav Hertz found that when electron projectiles hit atoms, the atoms absorb only very specific amounts of energy from the impinging electrons. Again this discreteness of energy arises, this time inside the atom. The electrons going around the nucleus are allowed only very specific energies, in complete contrast to satellites, which can take any orbit around Earth with any energy.

We actually owe our life to this discrete feature of the atom, because according to the regular laws of physics, a charged particle going in circles should emit radiation. This means it loses energy (in fact, this is the main drawback in circular particle accelerators). But if an electron would do that in an atom, it would continuously get closer to the nucleus until they eventually crash into each other, and if that were to happen, there would be no atomic structure, no chemistry, no biology, and no life. This does not happen because the electron going around the nucleus is allowed only these discrete energies; it is not allowed to lose energy continuously, which is what would happen if it were to radiate. It is only allowed to make energy jumps between allowed levels. Of course, Max, Gustav, James, and all the quantum gang that followed got Nobel Prizes.

Then came the photoelectric effect and other experiments which confirmed that, as de Broglie had suggested, all particles are also waves, and all waves are also particles. By the mid-1920s, there were already theoretical formulations of the theory, like the Schrödinger equation and the Heisenberg formalism. It was the first time humankind was able to explain the observed absorption and emission spectrum of atoms. Many other people, such as the Danish Niels Bohr, also made significant contributions.

Quantum theory was so strange that many of the inventors of the theory hated it. For example, to exhibit his dislike of the inherent element of randomness, Einstein made the famous statement that God does not play with dice. In fact, until his death in 1955, he continued to try to disprove the theory. In 1935 he published a famous paper in which he and colleagues claimed that the theory was incomplete. De Broglie hated the theory to such an extent that he refused to teach it in the university. He continuously tried (followed by David Bohm with a theory called Bohmian mechanics[19]) to find an alternative theory in which a particle is always an element of reality (i.e., exists), even when no measurement is being done. Another problem with the theory is the superposition principle, which states that an object can be in two states at the same time. For example, a single object can be in two separate locations simultaneously (see our amusing breezy video*). Schrödinger was horrified by the idea of the superposition principle, so he invented the story of a cat that is alive and dead at the same time. He wrote to his friend Bohr that

---

*Indeed, in many quantum labs around the world, including ours (https://tzin.bgu.ac.il/atom chip), we observe spatial quantum superpositions, whereby single quantum particles such as atoms and molecules, or even atomic clocks, are at two different locations at the same time. This state is typically realized in a device called an interferometer. For a description of a spatial superposition of an atomic clock, see our amusing popular-level animated video: https://www.youtube.com/watch?v=7yMTIO2gDfI.

he is sorry he ever had anything to do with the theory. Even today, artists still love to paint that poor cat, dead and alive at the same time.

Since in reality we never see superpositions in our macroscopic world, and in fact we only observe one state, the superposition principle requires us to imagine in addition some kind of collapse, whereby when we measure a system with an instrument in the lab or with our eyes, the superposition (of two, or even many, choices) collapses into only one choice. But what is the difference between the decoherence process we have mentioned previously and the collapse process? While decoherence kills the quantum-state coherence (so that, for example, an interference pattern cannot be formed in the double-slit experiment*), it still leaves unresolved the issue of which of the quantum possibilities will be eventually realized and observed by us. The collapse postulate takes care of that. It simply states that once an observation is made on a quantum system, our measurement chooses only one option. However, just like with the photon impinging on a partially reflecting piece of glass, we cannot know in advance which of the options the system will collapse into. In this sense it's not an informative postulate at all.

---

*The double-slit experiment is a paradigm of quantum mechanics. Feynman is known to have said that it in fact holds the only mystery. It served as the basis for fascinating discussions between Einstein and Bohr on the foundations of quantum theory. In brief, a single particle is sent toward an impenetrable wall that has only two slits, parallel to each other, through which the particle can traverse the wall. The particle is aimed at the wall, not very accurately, so that it can go through either slit with the same probability. Behind the wall there is a screen. Its plane is parallel to that of the wall. The screen is able to detect when and where the particle hits it. However, it can give us no information regarding though which slit the particle traveled. The particle can of course get to the screen only through the slits, typically through only one of them. However, if it is prepared in a quantum superposition so that it exists simultaneously in front of both slits, it may pass through both of them at the same time. Accumulating many such particles (sent one at a time) that were prepared in a quantum superposition and then hit the screen, one finds that while each particle hit the screen at a well-defined location, the distribution of the entire ensemble of particles creates spatial oscillations on the screen in which some locations on the screen received many particles and some almost none. This is called an interference pattern. The process of decoherence spoils the state of superposition, and the screen will show no interference pattern but rather two spots of particle concentrations directly behind the two slits, as intuitively expected. If we put a detector next to the two slits, which can detect through which slit the particle went, the superposition will collapse, we will know through which slit the particle went on an event-by-event (or particle-by-particle) basis, and again the accumulated signal at the screen, after many particles have gone through, will show no interference pattern but rather two particle concentrations behind the two slits. So collapse, like decoherence, kills the quantum nature of the system, but in addition we have (which-state or which-path) information provided by our measurement device. In the professional jargon, it is stated that each measuring device has its eigenstates, namely the type of system states it can resolve, and when measured, the system will collapse into one of these eigenstates of the detector. An interesting debate is whether collapse can also be induced by nature and not just by human measuring devices. See the next section for objective collapse models.

This collapse nevertheless stands at the base of the connection between quantum theory and the real world we observe, in which particles have well-defined features such as position and momentum. Collapse is consequently extremely important, and yet it is still external to the theory. It is an additional ad hoc postulate that was added to quantum theory.*

Actually, some physicists dislike this idea of the collapse so vehemently that they invented a many-worlds hypothesis, whereby all possible projections of reality exist, not just the one projected (collapsed) onto our brain at this moment, so the cat is alive in one universe and dead in the other.†‡ No one knows where these universes exist, perhaps in other dimensions. Quantum theory is so weird that it continuously gives rise to new ways of looking at it.[20]

So, the theory is very strange in the fact that it allows for objects to be in several places or states simultaneously, and also in the fact that it is statistical; namely, repeating an experiment with the same initial conditions can lead to different results—just like when a photon hits a piece of partially reflecting glass.

However, the strangeness of the theory does not end there. It also states that we cannot know the values of certain parameters with infinite accuracy, no matter how good our measuring devices may be. This is called the uncertainty principle, and if it is dominant in the brain, it inhibits to a large extent the possibility of a future supercomputer following the brain at the individual particle level, as the initial conditions would not be known accurately enough. (However, as noted, if decisions are made at a much higher level of clusters of neurons, in which the uncertainty principle is not a dominant effect, then the idea of the supercomputer simulating the brain

---

*The fact that quantum theory needs a postulate that is external to the theory in order to connect the theory to our observations is called the "measurement problem." One should remember that the process of collapse chooses a specific outcome, but it does so in a completely random way according to the statistics determined by quantum theory. For an advanced theory of collapse, see for example quantum Darwinism by Wojciech Zurek, and the next section.

†The many-worlds hypothesis avoids the need for the collapse postulate, as it stipulates that each possibility is indeed realized in its own unique universe or world. For those interested in the many-worlds interpretation, you may want to read the entry in the *Stanford Encyclopedia of Philosophy* (the entry was written by Lev Vaidman in 2021).

‡As a rather wild aside, I note here that some have already hypothesized concerning the many-world interpretation that hallucinations plaguing people with schizophrenia stem from intrusions of other worlds into ours; S. Tarlaci, "Quantum Neurobiological View to Mental Health Problems and Biological Psychiatry," *Journal of Psychopathology* 25 (2019): 70–84.

is back on the table. Furthermore, some claim that the uncertainty principle only reflects our knowledge of nature and not the state of nature itself, so it may be that even if the supercomputer cannot follow the brain processes, they are still deterministic.)

Another strange feature of quantum theory is perhaps best represented by the famous question: Is the moon there when no one is looking?[21] It's a zany sentence that changed the way physicists think. It tells you that quantum systems are so delicate that our measurements actually affect the results, so that we are altering the nature we want to observe. We can no longer be objective observers. But it also tells you something much more profound. It claims that there is no reality until a measurement is done. Indeed, scientists like Anton Zeilinger and colleagues have proven that even the mere supposition that reality exists independent of measurement leads to paradoxes.[22] Similar sentences describe the same mind-boggling thought, such as, "If a tree in the forest falls and there is no one there to hear it, does it make a sound?" This is not a word game. It is the quantum reality we observe in our laboratories. In other words, our measurement induces a collapse, and the collapse creates reality.

Perhaps the weirdest feature of quantum mechanics is the idea of entanglement.[23] Two particles that once interacted when they were at the same place at the same time, can now be galaxies away from each other but still have some ability to instantaneously affect one another (namely, faster than the speed of light). If such entanglement, as is utilized in the quantum computer, persists in the human brain, this may allow for complex parallel brain processes. However, one must realize that, at least as far as we know, such entangled states are extremely delicate and tend to quickly break, so decoherence processes in the brain would work even faster on these states. It is worth noting that entanglement between particles inside the skull and particles outside of it (even far away, outside the light cone) would prevent any supercomputer from predicting the brain, as not all of the relevant information would be at the supercomputer's disposal because particles from far away may be in a position to affect our brain. However, although some people claim that all particles in the universe are entangled in one way or another, in the laboratory we see that this natural universal entanglement is rather weak. We observe this weakness, since otherwise we would not have been able to do experiments on isolated systems in the laboratory and view them behaving just as we would expect isolated systems to behave according to theory.

I end this brief overview of the strangeness of quantum theory by recalling that it was Feynman, one of the great theoreticians of quantum theory, who said, "I think I can safely say that nobody understands quantum mechanics."

With your permission, I would now like to dive a bit deeper into the crucial issue of decoherence, as with strong decoherence there is no quantum mind. As noted, decoherence is the opposite of coherence, where coherence means a pure quantum state or process. Over the past century, since the birth of quantum mechanics, scientists have found that the main cause of decoherence is the environment external to our quantum system. The immediate environment of any system on Earth typically consists of many particles at room temperature. For example, in my lab, the quantum system is an atom in the middle of a vacuum chamber, centimeters away from its walls, and the environment is the vacuum chamber itself, which emits, for example, blackbody radiation, which interacts with the single atom. Such noisy radiation undoes the delicate quantum state. Additional forms of decohering radiation include electromagnetic waves from phones or power cables, and also light from the lamps in a room.

In the vacuum chamber, the atom is quite isolated (e.g., from air molecules hitting it), and most of the interaction with the noisy environment is suppressed, but in the brain, each particle continuously interacts with all the particles around it, which are just Angstroms away.* This means that the interaction with the environment is extremely strong. In addition, the brain is relatively hot, and this means that the particles are moving a lot, adding to the noise of the environment.

But even without the external environment, it is believed by many that large objects cannot display pure quantum behavior, as they are built from many particles (*degrees of freedom* in the professional language), and some sort of internal decoherence and collapse exists.† Be that as it may, the brain has so many particles that each neuron or even cluster of neurons has a huge

---

*The Angstrom (Å) is a unit of length equal to 10^{-10} meters. This is just about the size of the hydrogen atom.

†On decoherence independent of the external environment, but still within quantum theory, see, for example, my papers with Yonathan Japha and Carsten Henkel, *Physical Review Letters* 130 (2023): 113603, and *Physical Review A* 110 (2024): 042221. See also Section 7.3 on new physics below.

external environment within the skull, and consequently the standard decoherence of the fundamental building blocks of information processing in the brain—induced by the external environment—should be large.

So according to what we know today, after one hundred years of studying quantum mechanics, the brain should not enable quantum processes on a large scale (beyond single atoms or molecules)—a scale big enough to affect thinking processes. However, as noted, the door is still open to novel mechanisms that most scientists have yet to take into account. Indeed, some hints exist that nature has found ways to maintain large-scale coherence, as in the photosynthesis process[24] or, as noted, in the retina of birds, allowing them to "see" the magnetic field of Earth in order to navigate. This is part of an emerging but still speculative field of quantum biology.[25] For a general review of works suggesting that quantum mechanics plays a dominant role in the brain, see the notes.[26]

Another hint we may want to consider comes from general anesthesia. As there exists a wide range of chemicals with anesthetic capabilities, each of them very different from the other in their structure and interactions with brain matter, it is suggested that one should look at what is common to them, and this is perhaps quantum spin (the property of angular momentum—associated with rotation, and the magnetization that follows, originating within elementary particles such as the electron, proton, or neutron). Luca Turin found that under the influence of xenon, the simplest of all the anesthetics, fruit flies showed an increase in electron spin as measured through the use of electron spin resonance (though the origin of the signal is still debatable).[27] As the spin of nuclei has very long coherence times even in noisy environments, perhaps it is these spins that affect signal transport in the brain giving rise to consciousness. It should be noted that impressive work has already been done showing how quantum operations could be performed with spins in complex biological molecules.[28] I must say that I am quite excited by this prospect, as spin is one of the topics of my own work. (More about chemicals with anesthetic capabilities in the next section.)

There are in fact other hints as well telling us that in some cases decoherence may be suppressed. For example, a contamination atom in a diamond (sometimes called a color center, as these contaminations give the diamond its color), namely a different atom placed inside the lattice of carbon atoms that constitutes the diamond, shows long coherence times even at room temperature, and even though, as in the brain, other particles (the environment) are only Angstroms away. Also here, the coherence times of the nucleus are found to be much longer. In fact, the strong isolation of the nucleus from external disturbances is what is driving scientists to try and move from atomic clocks based on electronic spin states to atomic clocks based on the spin state of the nucleus (specifically, they are working with thorium 229). So, again, in some unique systems, decoherence is suppressed.

Furthermore, there is the concept of a quantum eraser, which is perhaps able to erase the information stored in the environment around the quantum system and in this way suppress decoherence. This information is produced as the environment interacts with the system, and it is this information that gives rise to decoherence. It could be that nature found a way to efficiently erase the information in ambient conditions.

A fascinating anecdote to keep our eyes on lies in the fact that people have started to use quantum computers to model the decision making of the brain. In 2023, the journal *Entropy* published a special issue titled "Quantum Decision-Making: From Cognitive Psychology and Social Science to Artificial Intelligent Systems." An example of an article appearing there is from the quantum computing company IonQ titled "Quantum Circuit Components for Cognitive Decision-Making." Is there some deep underlying reason that quantum circuits would be good at emulating human decision making?

Quantum theory presents many mind-blowing options for the future, and I limit myself to delving into one last option, which to me currently seems to be the most feasible of the quantum options related to our brain. Even if there is no suppression or bypassing of decoherence in the brain, could it be that from time to time the randomness of quantum processes at the level of a single electron, ion, atom, photon, or molecule are imprinted into a classical non-quantum signal, which may then be enhanced and preserved so that it may affect the brain? Could it be that evolution found a way not to fight or trick decoherence but to preserve true randomness? Namely, is it possible that quantum randomness is not averaged out in the brain like it is in a spaceship, which we need to be very deterministic? This would amount to having a true random number generator in our brain. Indeed, the commercial quantum random number generators already available on the market do just that. They take the randomness at the quantum level, protect it from averaging, and preserve and amplify it until it can serve classical systems. The brain obviously has many local processes of quantum randomness at the level of single particles, as noted above. In fact, the simple truly random process described at the beginning of this section, of a photon being transmitted or reflected by a piece of glass, also occurs for other types of particles, whereby different processes in the brain could imitate the function of the piece of glass. The big question is whether or not this randomness is then averaged out. As present-day, human-made, quantum-based true random number generators are quite simple, it is reasonable to speculate that evolution also found such mechanisms. Chaos, which provides very different final states for minute changes in initial states, may be part of such an amplifier in the brain.* If this is the case, then it is enough to make a huge impact, as described through the two-layer model of the brain presented in Chapter 8. In short, it would enable us to reach the right balance between determinism and randomness, thus achieving the requirement for life, namely of being *unpredictable but not random.*

Lastly, Eve, I would like to briefly go back to your original question, "Is the atom alive?" We have already noted, when discussing free will, that Kochen and Conway came up with a free will theorem which states that if scientists have some free will in the sense that their thoughts are not determined by the past (namely, by some Laplace demon), then elementary particles such as the atom should have the same free will. This was the first hint. But now, in addition, we have another similar hint. If it is quantum randomness that allows us life by putting us in the middle between deterministic predictability and randomness, thus making us *unpredictable but not random*, it seems the atom is alive in the same sense, as at some (statistical) level it is predictable,

---

*As noted, even without quantum processes, some scientists believe there is evidence that randomness exists in the brain (stochastic processes that are inherent to the brain), and furthermore that it plays a key role (see previous section on noise).

for example, through the equations of Claude Cohen-Tannoudji, whom we mentioned, but on another level, specifically on a single-measurement level, or a single-event level, it is governed by the randomness of quantum mechanics. Chapter 8 provides a two-layer model of the brain describing this tension within each of our minds between predictable and unpredictable.

As a preface to the next section dealing with new physics, one may ask if, scientifically speaking, quantum theory is the end of the road. After all, quantum theory has been proven right in thousands of experiments. However, the history of science, and furthermore our belief in the infinity of reality, should make us deduce that quantum theory is not the end of the road and that at the very least it will be extended, just like Einstein's relativity extended Newtonian physics to a more accurate description.

It is interesting to note that there are many no-go theorems concerning quantum mechanics which state what a future theory replacing or extending quantum mechanics cannot be. However, as I already noted, one of the greatest theoreticians of quantum mechanics, John Bell, said that all these theorems prove is our lack of imagination.[29]

In fact, Bell even said something that is stronger than that. He said that quantum theory carries in itself the seeds of its own destruction. In the next section, I speculate regarding the possibility of new physics.

3. NEW PHYSICS

New physics may of course be necessary for any future understanding of how the brain works and what ultimate determinism its clockwork adheres to. Consequently, it is clearly warranted to try and briefly contemplate the possibility of new physics.

Merely looking at the history of scientific revolutions and taking account of our belief in the infinite complexity of reality, and furthermore, realizing that we humans have been studying nature for such a relatively short time, it stands to reason that much more new physics has yet to be discovered. One is on safe ground stating that what we do currently know, sometimes referred to as the standard model, is a mere fraction of what we will know in the future.

I have already noted that some physicists hold the view that "there's no reason to be agnostic about ideas that are dramatically incompatible with everything we know about modern science."[30] On the other hand, I previously quoted John Bell as saying that "proofs of what is impossible often demonstrate little more than their authors' own lack of imagination."[31] I personally tend to side with the latter point of view, although we should all be on the guard, as loosening our scientific rigor will clearly allow endless expressions of superstition and charlatanism.

I have already mentioned in passing some of the ideas and consequences of new physics, but now let me present these thoughts, as well as new ones, in a more detailed manner.

But first, perhaps I should acknowledge the very impressive efforts scientists have been making and are making to discover new physics. It is not at all a trivial feat to go against a paradigm according to which you have been trained your whole life, to the point that it has become your intellectual comfort zone. More so, the structures of scientific activity, including promotion and funding committees, are typically based on senior scientists looking at the ideas of younger scientists, and this can create friction and barriers when new ideas are suggested. I recall a discussion I had over coffee with my good friend, Nobel laureate Theodor (Ted) Hänsch, who said that the demand made by the establishment that young scientists publish papers quite frequently in order to be promoted is not allowing them to evolve into deep thinkers taking on high-risk, high-gain, long-term ventures. He made a beautiful analogue by calling this barrier young scientists are facing the Zeno effect. In physics, the Zeno effect is an effect in which when you measure a quantum system very frequently, it gets stuck in the same state and cannot evolve to other states. Indeed, the original motto of universities, *Libertas Disputandi*, namely, the freedom to argue, to have opposing opinions, has been corroded due to the legitimate need of academic institutions to try to accurately assess the capabilities of its members during their younger years. The bottom line is that new physics is an extreme case of going against the flow. It's a daunting task to think of a new idea, not to mention the horribly hard work that is required in order to prove it is correct, and in addition there is the friction I described above. Scientists should therefore be commended for their strides in finding new physics.

Dear Eve, the rest of this section is made of details and contemplations concerning new and old physics. If you don't feel like it, please just go to the final paragraph starting with "To conclude . . . ."

Will new physics come from an experiment in the lab or a thought experiment? We of course don't know. Quantum theory was mainly born from unexplained results in experiments, while the theory of relativity was mainly born out of a thought experiment. In the lab, as we try out new experiments, perhaps we will stumble over new observations that can only be explained by new physics, such as theories requiring multiple new dimensions (e.g., string theory with at least ten dimensions). New observations can also come from improved technology in the lab, as we may probe new regimes we could never access before. For example, we are becoming sensitive to effects at shorter and shorter time and length scales. Some physicists claim that we will see new physics only at the Planck scale of 10^{-35} meters, but we should all hope it will happen at much larger scales than that.

When speaking of new physics, we should also acknowledge the fact that many open questions (namely, those yet unanswered) still exist in physics. For example, why does there seem to be much more matter than antimatter in the universe? This seems to contradict the symmetry of nature. And then there is the issue of the acceleration of the expansion of the universe. Acceleration requires force, and force requires energy. No one knows what the energy is that drives this acceleration in the expansion of the universe, and consequently it has been termed dark energy. Similarly, galaxies are observed to be turning too fast relative to the amount of material (stars) they hold, where both of these parameters may be measured with our telescopes. The gravitational pull of the mass we do see is not strong enough to keep the stars from escaping away from the galaxy due to the centrifugal force. The fact that the stars are not observed escaping means that there is much more mass there, mass that we can't observe, probably because it does not emit light. No one knows what the nature of this mysterious material is, and consequently it has been termed dark matter (spaceships in the future will have to take great care not to collide with this invisible stuff). The Wikipedia entry titled "List of Unsolved Problems in Physics" is quite extensive, and if you have some time I encourage you to take a look.

Concerning the quantum mind discussed in the previous section, I must say I was actually surprised not to find the quantum collapse postulate among the unsolved problems listed in the above-mentioned Wikipedia entry. Numerous scientists find this postulate to be unsatisfactory and have tried to replace it with additional theories, such as the many-worlds interpretation, which we have already mentioned, or with different objective (i.e., independent of human measurement or even existence) collapse models, such as the Ghirardi-Rimini-Weber (GRW) model, the continuous spontaneous localization (CSL) model, and the Diosi-Penrose model. The latter, related to gravitational effects,* is the motivation for numerous experiments, including an experiment conducted in my lab in which we are attempting to put a single massive object in a spatial quantum superposition, namely, in two different locations at the same time.[32] This is expected to create a superposition of space-time itself (as mass is a source of space-time curvature), and such a superposition is conjectured to be highly unstable.

Quantum collapse and the decoherence that precedes it are extremely relevant to any discussion of the brain. Although I briefly explained the difference between them in the previous section, since it is quite subtle, and because it may lead to new physics, I believe I should give more details. As noted, according to quantum theory, an object can be in several states at once, where a state can be a specific value of a measurable parameter like momentum, position, energy, or spin. If the object is indeed in such a pure quantum state, there is a strong mathematical relation between its different parts, which is called a phase relation. We then say that there is coherence (the opposite of decoherence), namely that the object is quantum. We can measure whether such a phase relation exists in an interference experiment. Let me take, for example, a particle that is simultaneously in two locations. Let's call them Left and Right. A pure quantum superposition state means that there is a phase relation between the two states (we refer to the two states as wave packets). When decoherence kicks in, this phase is destroyed, the

---

*See, for example, Sandro Donadi et al., "Underground Test of Gravity-Related Wave Function Collapse," *Nature Physics* 17 (2021): 74–78; Di Zheng et al., "Room Temperature Test of the Continuous Spontaneous Localization Model Using a Levitated Micro-oscillator," *Physical Review Research* 2 (2020): 013057.

delicate quantum state is destroyed, and an interference experiment (e.g., the double-slit experiment described previously) will give a different result than it would for the quantum state (specifically, no interference pattern). We are then left with a classical state (sometimes referred to as a statistical mixture) for which we simply have a one-half probability that the particle is in the Left position and a one-half probability that the particle is in the Right position, exactly like the heads-tails probabilities when we flip a coin. However, when we measure the particle position (e.g., place two detectors, one next to each slit), we will find only one of the options. This move from two options to one option is called collapse (sometimes also referred to as localization or reduction). I would consider the solving of the collapse problem as new physics, which, as we noted, may also have strong implications for the workings of the brain. I add here that quantum behavior is not only hypothesized to be affecting the brain. Some believe that the brain, or more precisely our consciousness, is what causes the collapse of the systems we observe.[33] The idea that consciousness leads to collapse goes back to Eugene Wigner, the Hungarian American theoretician.

I previously noted that calculations were done back in the 1990s stating that quantum effects cannot be dominant in the brain, but I have also conjectured that we may be unaware of subtle defense mechanisms that evolution has developed to safeguard quantum coherence. In fact, an advocate of the quantum mind might say that the big leap between the mind of apes and that of humans happened when evolution found a way to enable the quantum mind. This is obviously a far-reaching speculation, but nevertheless a thrilling one. As noted, the emerging field of quantum biology, looking into the extraordinary efficiency of photosynthesis or the apparent ability of birds to sense the Earth's magnetic field, may give us hints in this direction.[34] In addition, if room-temperature superconductivity is achieved, whereby long-range quantum effects survive the noisy room-temperature environment, this may also point in the same direction. Perhaps the rather exotic cytoskeletons and microtubules in the neurons (resembling tubes having a 14-nanometer [nm] diameter on the inside and 25 nm on the outside, and up to hundreds of micro-meters in length) may serve as such a hypothesized defense mechanism. Penrose and colleagues think they can,[35] although the decoherence rate, as I noted previously, is still under debate. While originally these microtubules were simply thought to have the task of transmitting neurotransmitters from their production site at the soma (body of the neuron) to the synapses, they may now be understood to also control synaptic changes,[36] which may even give rise to a computational role.*[37] Can they perhaps also enable long-range quantum effects? Currently, we simply don't know. It is a fascinating hypothesis.† What is clear is that to date, quite a few scientists are still convinced that quantum effects in the brain can be connected to these microtubules,[38] and recent work even draws on an anesthesia experiment.[39] Other scientists are also convinced that quantum theory indeed plays a major role in the machinery of the brain, and they have come up with various ideas on how this could be made possible. I have already mentioned the idea of connectivity by light. In another example, Matthew Fisher from UC Santa Barbara believes that the Posner molecule can protect quantum systems (a neural qubit) against decoherence and even give rise to entanglement.[40] He is specifically thinking about nuclear spins, which have long coherence times. We have already mentioned these spins in the previous section. In fact, recent work by others strengthens this hypothesis.‡ There are also those who wonder if the quantum geometric phase may be involved, whereby the quantum information processing may be so fast that standard decoherence does not have enough time to act. In any case, scientists are continuously wondering what quantum-type experiments

---

*By the way, if indeed there is computational power in the synapses, this may increase the computation rate of the brain of say 10^{14} per second (as there are about 10^{11} neurons and each one fires about one thousand times per second) by many orders of magnitude because both numbers may considerably increase.

†Personally, I think it could be a viable hypothesis because of the following rather technical argument. A trapped particle typically enters the quantum regime if its de Broglie wavelength ($\lambda = h/p$, where h is the Planck constant and p is the particle's momentum) is on the order of the size of the trap (or potential well that holds or guides it). The de Broglie wavelength of an electron at room temperature (say, T = 300 degrees Kelvin) can be calculated by noting that its velocity is something like the square root of KT/m, where K is the Boltzmann constant and m the mass of the particle. For an electron with a mass of about 10^{-30} kg, the velocity comes out to be about 5×10^4 meters per second. From the velocity, it's easy to get the momentum and the wavelength, which strangely enough comes out to be 10 nm, about the same size as the internal diameter of the microtubules! However, it's debatable whether there are free electrons in the brain. Charges are typically carried by atomic ions, and these, due to their large mass, have de Broglie wavelengths that are smaller by at least a factor of one hundred. Furthermore, along the long axis of the microtubules, the same calculation does not give a quantum effect, but still the transverse quantum state may somehow propagate. One should also note that Penrose goes beyond the mere existence of quantum states and hypothesizes that quantum computing of some sort takes place.

‡In this interesting work by principle investigators Yonatan Dubi and Lev Mourokh (published in M. Khonkhodzhaev et al., "Persistence of Correlations in Neurotransmitter Transport through the Synaptic Cleft," *Biology* 13 [2024]: 541), they calculate that correlations—a key ingredient of the quantum brain hypothesis—do manage to survive in the harsh conditions of the brain. Obviously, much more work needs to be done before we can completely accept the notion that quantum mechanics has dominant consequences at the decision-making level of the brain. This finding goes well with my previous statement that earlier calculations of decoherence in the brain, showing that the quantum mind cannot be a valid hypothesis, were limited to very specific processes and in no way covered the full spectrum of conjectured brain processes.

we could already be making in order to learn new things about the quantum-mind interface,[41] as well as what kind of quantum-based sensors we could be employing to find quantum effects in living neural networks.[42]

Decoherence rates are a crucial component in this story. If we find that we simply did not know how to properly estimate the rate of decoherence in the brain, that will be big news, but still not new physics in the sense that no new laws of nature are invoked, including, for example, new particles and new forces, or that the space we live in has more than four dimensions. Similarly, if we find that the weak intermolecular forces acting over relatively long distances named van der Waals (vdW) or spins play a dominant role in the brain, this will be huge news, but not new physics in the above sense. The hypothesis that vdW forces or spins, or perhaps even chirality (the way molecules are twisted*), could play a significant role may find support by analyzing the effect of the wide range of chemicals that exhibit anesthetic capabilities, such as chloroform ($CHCl_3$), nitrous oxide (N_2O), halothane ($CF_3CHClBr$), ether ($CH_3CH_2OCH_2CH_3$), isoflurane ($CHF_2OCHClCF_3$), ketamine, and xenon. As their chemical reactions with the brain must be widely varying, we should try to find what they may have in common that allows them all to have anesthetic capabilities. This may be their electrical polarity (electric dipole affecting vdW) or their spin,[43,44] or other degrees of freedom. For recent studies on the effect of anesthesia, see references in the notes.[45]

So what new physics can we imagine? There are numerous ideas going around. At this stage, they may be termed myths because they still don't offer any predictions that can, at this time, be empirically tested. These include the many-worlds theory noted previously, or string theory with its many additional dimensions. Any empirical evidence showing that this sort of idea is correct would constitute new physics. Another such idea claims that the universe has some form of consciousness and that this is the real explanation for why the fundamental constants of nature, of which there are dozens, such as the mass or the charge of the electron, are exactly as they are. A tiny change in these numbers would not have enabled life as we know it. In this idea

---

*Chirality is best described by the way screws turn, right-handed or left-handed. For a machine, it would not matter if the screws are right-handed or left-handed, but for us humans to live, as Louis Pasteur found out, the same molecule can be an essential element of life with one chirality while a deadly poison with the other.

promoted by Philip Goff in his book *Why? The Purpose of the Universe*, the universe tuned itself to have life. This seems to suggest something that goes beyond the materialistic view that physics currently holds. He is not suggesting some sort of self-aware consciousness that is similar to how religions view God, but also not the atheism à la Richard Dawkins. In my view, it could very well be that there is some hidden variable or physical parameter, which, as I noted previously, we are still not aware of, and for which the creation of intelligent life is the solution for some equation, and this forces the fundamental constants to be as they are. But again, there is currently not a shred of evidence that this is the case.

New physics can also come out of any one of the open questions mentioned above. For example, could dark matter or dark energy, which are now only thought of in models of the universe, also bring about some new mechanisms in the functioning of the brain? An example of a hypothesis bringing to the table completely new physics could be a new model of particles and interactions that allows for true randomness without suffering from decoherence. This would result in distancing the brain from being a deterministic machine. Another example is the idea that at the base of consciousness lies a new form of physical law that is noncomputable.[46] Such a theory of physics has never been encountered by scientists. This gives rise to a new option in which even if nature and our brain are deterministic, the brain is still unpredictable by an external agent because the external agent with its enormous computing power still cannot compute and predict the brain. Namely, we are a machine, but according to our definition of life, a living machine!

For example, Penrose argues that at the interface of quantum mechanics and gravity may lie a new theory that is mathematically noncomputable. Just like in computer science some problems have been hypothesized to take an exponentially long time, not to say to be unsolvable for all practical purposes (FAPP).[47] A specific example is the noncomputability of the topological equivalence problem of four-manifolds, as found by Robert Geroch and James Hartle (1986).[48] As quantum gravity includes superpositions of four-dimensional objects, it may include noncomputable elements. The three-body system mentioned earlier is a simple example of computational complexity, whereby there is no general solution that can be expressed in terms of a finite number of standard mathematical operations.[49]

It is hard to reconcile a physics theory with a theory that is noncomputable. It sounds like an oxymoron. However, recalling what we already know about the determinism of quantum mechanics at the statistical level, meaning it's completely computable and therefore predictable, while being completely unpredictable at the single-event level, we may say that quantum mechanics gives us a hint of how such new physics may play out; namely, it teaches us that such weird new physics is possible.

Another possibility for completely new physics would be if we found out that the mind and the physical world are completely separate entities and that the physical world may be affected by the mind but not the other way around. Descartes, for example, hypothesized that "the mind is outside of regular physics and intervenes on the physical world." This opens the road for some sort of free will in a deterministic world, and some physicists are even thinking of ways to test it. It goes without saying that there is still no proof whatsoever or even the slightest hint of something of that sort.

Finally, some scientists, such as Yakir Aharonov from Tel Aviv University, hypothesize that we will encounter new physics in a way that is relevant to the brain, as this new physics will be born out of effects having to do with complexity. Similar to the concept of emergence, which we discussed earlier, these scientists conjecture that we may find that above some threshold of complexity, our existing physical laws cannot describe nature. Again, there is no proof.

To conclude this brief description of the idea of new physics, I would like, in this new context, to emphasize once again the main point of this book: The powerful potency of our new observable P is that we don't need to wait for new physics—just as we don't need to wait for the scientific community to agree on a quantifiable definition of consciousness—in order to learn something extremely fundamental about who we are. These searches for new physics and for consciousness are very important indeed, but they are expected to take a long time. With the help of our new observable P, we can already gain profound insight into who we are and what we may become.

We are now ready to visualize the battle taking place in our brain between random and deterministic, which, according to our definition of life, is nothing short of being the most important battle we will ever fight.

8

The Bottom Line of Predictability

1. A SIMPLIFIED MODEL FOR HOW OUR BRAIN PRODUCES OUTCOMES

Dear Eve, this book, my answer to you, is all about who we are in a very pragmatic manner that may be tested. It focuses on the measured output of the brain in terms of actions we make, specifically measuring predictability, in order to avoid all those vague terms that have dominated the philosophy of who we are for centuries, without providing closure, or even some signs of convergence, like *consciousness*, *self-awareness*, *soul*, and *free will*. Although we mainly focus on external actions, it is insightful to build a simplified model of how the brain produces will, which is then followed by action. This simple model should, on the one hand, be consistent with everything that is currently known about the brain, and on the other, it should provide some insight regarding the conclusions we are about to reach. It will of course be presumptuous to try to distill the humongous body of work that has been done on the brain into a simple model, but such a model is warranted for the purpose of this book, namely, in order to visualize how different brains may be found to be situated at different points along the axis between nonpredictable (random) and predictable.

Our simple model is built from two brain layers, a random number generator and a filter, which we will directly connect to the two extreme results of the experiment now underway, or the two extremes of our definition of life, complete randomness and complete predictability. In the section on noise, we have seen how cognitive scientists emphasize its dominant role in the function of our brain. Some even postulate that it is the source of our free thought and free behavior. As noise is for all practical purposes random, one may therefore assume that a random seed stands at the deep origin of every decision we take. It may in fact be the mechanism responsible for allowing our brain to scan all possible actions. This is perhaps consistent with the findings of the Libet-type experiments, which showed that decisions are made by the subconscious before we are made aware of them in our conscious mind, where here we treat the very core of our subconscious as a random number generator.

But why then are our actions not random? A photon impinging on a piece of glass will sometimes be reflected off the glass and sometimes be transmitted through it. Quantum theory tells us that this is truly random, but our human behavior seems to be the exact opposite. Most of our decisions seem to be well motivated by the wish to achieve a certain goal. Indeed, throughout this book we have been claiming that we may even be deterministic to the point of being completely predictable.

We thus introduce the second brain layer, which is the filter. We recall that Libet and his followers introduced the notion of free won't. This is a veto right that the conscious mind has over the will produced by the subconscious. We indeed don't do many of the things we want to do. Here we identify this veto right as a filter, which is in charge of allowing only some of the random numbers or thoughts to pass through and turn into action. For example, if you are in a shop and you see something nice but you can't afford it, your random layer may also suggest, among several options it allows you to explore, that you steal it, but your filters will stop you. These filters may be due to fear of getting caught or to moral and ethical considerations. For example, this latter filter may have been established in your brain when you once heard in class Immanuel Kant's famous directive that the principles governing your actions, namely the individual's actions, must be such that all individuals may live by the same principles, and it resonated with you. The functional explanation of the filter layer may also be aligned with other psychological theories, such as Freud's id, superego, and ego. In this case, the filter blocking the wish to steal is connected to the superego.

I also recall that in Chapter 5, I described how Thorndike explained that the mind repeats what works for it. Responses that produce a satisfying effect in a particular situation become more likely to occur again in that situation. This may be exactly described by the forming of filters, whereby a satisfying result produces a filter that would make this course of action much more likely in the future. Such a forming of a filter based on past experiences may be similar to what is called a target function in the field of AI, whereby the training of AI on data sets that include input and output allows it to make predictions on future unknown outputs and to aim for the best result. We noted in Chapter 5 that, in this sense, Darwinism can simply be looked at as a survival-of-the-fittest feedback loop, but now we could say that Darwinism can simply be looked at as a survival-of-the-fittest filter system.

Our filters are extremely complex. They are our personality traits. They are made, on the one hand, by the genetic makeup of our brain and, on the other, by the links formed in our brain following years of all kinds of education and learning processes. They also need to be able to process and adjust immediate inputs from our senses. These well-established filters may be in charge of very complex vetoes, such as which career to choose or which partner to marry, and they can also be in charge of our immediate feedback loop helping us survive the moment. For example, if you are on the edge of a cliff, and deep in your subconscious there is a thought to try to imitate Superman by jumping off the cliff, your survival filters will immediately kick in and kill the thought. Of course, a completely different filter may have brought you to the cliff that morning. For example, because your previous visits to the cliff were pleasing (you liked the view and the wind), and because the weather is currently nice, your random thought after breakfast of going for a stroll to the cliff was allowed to be turned into action.*

Some filters may have an unconscious (automatic) nature, and they may occur in the 500 millisecond (ms) delay between when the readiness potential is measured and the time at which we report that we recognize a will. Other filters may be applied in our conscious mind processes, specifically in the 200 ms between the time that we recognize a will† and the actual motor response

---

*We are of course ignoring here the complexity of how the brain chooses which filter to engage for which situation.

†The time of the will in Libet's experiments was a rather subjective measurement reported by the person himself, by stating the time he/she saw on the clock when they felt they had a will to push their finger. On average, 200 ms later, the finger was indeed pushed, and a click was made on the computer keyboard. This is perhaps a reasonably accurate statement as the visual perception temporal resolution is around 30 ms.

(the time in which a command reaches the body once it has left the brain may be as fast as 10 ms). Namely, we know that we want to act, but then, due to some considerations for and against, we veto the action.

Our filters may be too weak, in which case we may be impulsive and even violent, and they may be too strong, in which case we may be too shy or possess a nature that succumbs to hierarchy and authority. It is quite fascinating to consider that some filters must survive for decades if they are, for example, to navigate our decision making toward a long-term goal such as becoming an excellent pianist or a highly qualified medical doctor. Some filters may also return an inconclusive answer regarding a specific action (e.g., asking someone to marry you), in which case the filters will request further input, which could mean, for example, more dates with a specific woman or man before the filters can be conclusive. Any random thoughts about dating other people may then be blocked until the request for further input concerning a specific woman or man is completed.

Filters do not only affect our decisions. They also affect how we interpret reality, which in turn affects how we try and solve a problem. This subjective limited view of reality is sometimes called cognitive fixedness.

Our two-layer model, in which one layer is responsible for randomness (unpredictability) and the other for predictable determinism, is in a way reminiscent of what David Bohm tried to do with his new interpretation of quantum theory. He tried to find an interpretation in which a particle is still a particle (as we thought of it before wave-particle duality appeared; see Section 3.3) and everything is deterministic (i.e., the particle follows a well-defined trajectory). To retain the random nature of quantum theory, he had to hypothesize that the source of the particle, the origin where it is decided which of the many well-defined trajectories it will take, is statistical in nature, namely, random. Likewise, our source, the core of our thoughts in our two-layer model, is random noise. The rest, namely our filters, is deterministic.*

---

*Indeed, the two-layer model presented in this book is reminiscent of other points of view or descriptions of the mind. For example, in Rafael Malach's 2024 article "The Neuronal Basis for Human Creativity," the mind is described as a "unique integration of random, neuronal noise on the one hand with individually specified, deterministic information acquired through learning, expertise training, and hereditary traits."

In Chapter 1, I noted that the near-future measurement of predictability concerns most if not all of our actions that take place in goal-oriented situations in which we perform some task to achieve a practical goal. Most of our actions are of this type. However, one may create artificial tasks that are not of this type, such as asking a person to think of a random number. This is a task that has an inherently random nature where the only goal is to be as random as possible. Just as our brain has the ability to engage different filters for different situations, the brain may be able to give dominance to randomness when required, as in this task. This means completely suppressing the filter layer.

The above mechanistic model of the brain does not have free will in the usual sense, as there is no freedom to speak of. So, if we reflect philosophically for a moment, we may wonder about the justification for punishment. This is exactly the kind of question religious scholars like Maimonides, whom we have already met, asked. Or similarly, as we have previously asked, "Can a person be morally responsible for her behavior if that behavior can be explained solely by reference to physical states of the universe and the laws governing changes in those physical states, or solely by reference to the existence of a sovereign God who guides the world along a divinely ordained path?" The answer is that, according to our simple two-layer model, punishment is extremely important for our society, as it is part of the learning process that forges the filters in each of us. So, while it may not be ethically justified, as there is no free will, it is a necessity of society both in forging the filters in the minds of humans and, furthermore, in simply getting humans who have bad filters off the streets.

But the most important aspect of our model, and the justification for its existence, is that it is able to directly connect to our new definition of life, *unpredictable but not random*, or in quantitative language, P = 0.5, namely, halfway between the extreme of randomness and the extreme of predictable determinism. The two-layer model explains how different brains may be found to have a different level of predictability, and it also reflects our definition for life, as one layer is random and the other deterministic. The correct interplay and balance between the two may thus enable us to reach the optimal P = 0.5. This understanding will allow us in Chapter 9 to investigate the possibility of a pill that could cure us and to concoct some cookbook recipe for making it.

2. CONCLUSION

Eve, before I go on, as you requested, to discuss the topic of human life in the language of the humanities, allow me to now conclude the scientific part of the book. Following the shocking, perhaps even outrageous, words spoken by you in my lab, asking whether the atoms I communicate with are alive, I was dismayed, and in fact I faced a never-ending chain of thoughts haunting me through the course of many sleepless nights. For if the atoms are indeed dead, as I always thought they were, could it be that we are dead as well? For we are nothing but our brain, and our brain is nothing but atoms.

To answer your question, it was clear that I had to first define what we mean by *life, human life*, as you agreed that what you meant by the word *life* in your question about atoms was in fact human life. Namely, this was the ruler against which you wanted to measure the atoms. So, what is the definition of *human life*? I first made a deep dive into existing theories of consciousness meant to solve this exact enigma. But I soon found that if I am to provide a reasonable answer that could give some clear-cut bottom line within our lifetime, I must go beyond consciousness, for it is still very much ill defined. Indeed, as evident from the scientific field of consciousness studies, consciousness is extremely complex, and at least for the foreseeable future, it stands to reason that there will not be agreement on how it is created or how to quantify it, not to mention agreeing on at what level of consciousness we enter the human standard of life. It became evident that without the bold move of going beyond consciousness, any answer I send you would simply constitute more of the same.

I therefore had to rely on a new observable, and I found that predictability may serve as such an observable. Furthermore, to my surprise, I found that there is already an ongoing experiment to measure it, an experiment that is now emerging as the most profound experiment ever conducted on humans. In this book, my answer to you, I described how this high-impact observable could shed new light on who we are within our lifetime.

Thanks to your question and your demand for an accurate answer, and by utilizing a well-defined measurable observable, predictability P, as a parameter that provides insight concerning the most fundamental questions about who we are, we have in fact unveiled a complementary approach to yet unclear concepts that have dominated the philosophical discourse over the years, such as consciousness and self-awareness.

P, your predictability, will be measured by utilizing half of the data harvested regarding your personal choices so far in your life, what has been termed your intelligent history, to build a model of you by training the AI, and then the other half is compared against the AI's predictions of your decisions. This enables calculating a P value for each individual human.

The new observable is also potent because it may be measured by brain outcomes, namely, the actions of a person. Actions are a clear-cut data point, whereas internal brain processes, such as those measured in the studies of consciousness, are much more subtle and ambiguous.

Our new observable and the more traditional observable we call consciousness may very well be connected. On the one hand, many cognitive scientists would agree that mind-altering drugs induce a higher (or expanded) level of consciousness. On the other hand, many would also agree that mind-altering drugs induce divergent thought, namely, creativity, which is tantamount to less predictability. It is not by chance that so many artists find it helpful to use hallucinogens (see the introduction and the following section on creativity). It is therefore evident that our new observable may be directly connected to what we call consciousness.

As predictability is incompatible with free will, measuring high levels of predictability in humans in the near future will bring about earthshaking implications, specifically, that free will does not exist. Similarly, any other assumption of a free agent influencing our behavior, such as a free soul, will be inconsistent with the facts, if indeed a high level of predictability is measured.

We may even talk of life and death. We have shown that if we equate a high level of predictability with what we have termed machine death, this new observable allows us to go beyond the present rudimentary definition of the border between life and death—based on those flat lines on hospital monitors measuring the heart and general brain activity. Namely, even if the lines on these existing rudimentary monitors are not flat, the measurement of P may still tell us we are dead. Although we are speaking here of mental death, we use the word *death* not as some metaphor, as mental death is real death. If we wish to avoid randomness ($P = 0$) as well as to avoid being completely predictable ($P = 1$), we may define the utmost of human life, that which we should aspire to, as *unpredictable but not random* ($P = 0.5$).

Finally, I have conjectured, and this may soon be put to an empirical test, that in the not so distant future, instruments based on AI and Big Data will be able to predict the decisions and actions of some people with a very high level of accuracy. Obviously, the ability to predict us is now making its first baby steps, but the trajectory is clear, and what remains is to make a sound extrapolation. Perhaps the hardest thing for AI when it comes to predicting any future choice will be to understand the context surrounding a human when making a decision. However, the exponential growth of these capabilities seems to clearly suggest that within a short time, AI will also understand the human context. These capabilities include data harvesting (including wearable sensors), detailed human models (either through theories of personality types or collaborative filtering), and enormous computing power and memory (i.e., housing the entirety of the individual's intelligent history). These allow us to make an educated estimate that the context will also become clear to AI in the very near future. It also seems likely that in some people predictability will be measured to be high, while in others less so, so that predictability will be found to be a personality trait.

To try and estimate what values of P people will score in the near future, we must go deeper into models of the brain. Specifically, to anticipate that AI and Big Data will achieve ultimate predictability of humans, namely, P scores close to one, we must assume that the black box that we call the brain is deterministic.

Concerning the determinism of the brain, it seems there are several possibilities: First, that there is strong universal determinism, which means that all our thoughts are predetermined, and we are pure machines just because nature is a pure machine. The second option describes a strong determinism giving rise to a partially unpredictable brain, at least FAPP, due to noise playing a key role in our brain. While this is not free will, it may be called free behavior, and indeed some scientists are already using this term. But as we explained, no real freedom is involved. The third option states that classical (non-quantum) physics is still dominant on the decision-making scale of the brain; however, nevertheless, there is no universal determinism but rather a weak determinism that emanates from genetics, education (not to say brainwashing or indoctrination), and the never-ending aspiration of the chemical machinery of the brain to reach chemical balance. Here, there may be some freedom. Then there are additional options, which emanate from dominant quantum processes, and finally the last option takes into account the possibility of completely new physics.

Let me briefly further explore the possibilities mentioned above of strong and weak determinism, the main quantum scenarios, and the possibility of new physics.

In strong determinism, there is no freedom of choice; it simply does not exist. In this case, any test of predictability quantifies how predictable nature has made us in the eyes of (also predetermined) tests and probes. It would seem that the deterministic pathways of the evolution of the universe bring about the creation of higher and higher forms of the mind as more and more species are created, from the feedback loop of the lizard to that of humans.

But even in our own species, it may very well be that the deterministic universe has given rise to a wide spectrum of normalized predictability value P, as the volume of information that needs to be accumulated on different individuals before they are predictable with a high rate of success is most probably widely varying. This is quite an amazing scenario: If a brain that is closer to $P = 0.5$ is defined to be of an elevated level relative to a brain that is closer to $P = 1$, and if indeed evolution creates mechanisms to distance the brain from $P = 1$, as we discussed via our two-layer model in the previous section, then it seems we are witnessing a race between the evolution of unpredictability in the elevated biological mind and the predictive powers of Big Data and AI, all created by the same deterministic evolution and by the same deterministic mind.

In the case of strong determinism, it was also determined ever since the Big Bang that the day will come in which humans will be aware of their predictability, and even find ways to measure it.

Specifically to ways in which evolution is distancing us from $P = 1$, strong determinism may be rendered unpredictable FAPP if noise plays a key role in the brain. As we have seen, many systems in nature, as well as artificially made systems such as electronic circuits, show fluctuations that we call noise. Such fluctuations are a part of the deterministic universe, but they make predictions about the system much harder to attain, as many more degrees of freedom need to be taken into account. While this is clearly not freedom, it does move us away from the machine death of $P = 1$. As the effect of this noise is obviously limited by the constraints and boundaries formed by genetics and learning that make up our personality traits, and which we have termed filters, it may be that the door is open for us to increase the effect of noise in our mind. Namely, weakening the filters would allow more space for noise, which in turn would enable us to scan more options and be less predictable.

In the fortuitous situation of weak determinism, some freedom may be assumed to be possible by the underlying physical laws of nature. As we have seen, this is what the intuition of quite a few philosophers and scientists tells us. One may contemplate that in analogy to Pascal's wager,* it makes much more sense to believe in the existence of some freedom, since if you believe and it doesn't exist, you lose very little, and if you believe and it does exist, you win a life of transcendence. On the other hand, if you don't believe and it does exist, your losses are immeasurable. Consequently, even in the face of the omnipotent Laplace demon, we also consider here the possibility of weak determinism.

However, there is no good definition of what this freedom or free will means, and as in this book we have made it our mission to stick to words we can clearly define, we may speak of freedom only in the sense that there may be some actions stemming from within, namely, that are not fully determined by an external stimulus or instruction and may thus be called internally motivated actions. In any case, as predictability is the opposite of any kind of freedom, whatever its definition may be, we must strive to fight against our personal predictability. To do so, in the language of our two-layer model, also here (as with strong determinism) we need to find ways to increase the dominance of internal noise and decrease the strength of the filters.

Be the definition of the allowed freedom within this weak determinism as it may, we must always remember that any freedom is obviously limited by the great barriers of our rudimentary brain power, limited senses, ego, genetics, and education. The existence of this freedom is surely not a given, even according to the intuition of the most optimistic proponents of the idea of a free brain. It is a potential that can only be realized through hard work. What matters at the end of the day is not the allowed space of choices but the choices we eventually make and that we eventually act upon. Acquiring freedom, or at the very least unpredictability, and utilizing it are a manifestation of the elevated mind of a living human, at least according to our new definition of *life*.

---

*See footnote in Chapter 6.

And then there are the quantum options. I focus on the two main extremes. If quantum processes are dominant enough in our brain to affect our thoughts, the randomness of quantum mechanics may introduce true random noise into our deterministic mind (in contrast to the above classical noise). This is because quantum theory has an inherent random part built into it; that's why it's called a statistical theory. A specific set of initial conditions may lead to several possible outcomes. This randomness can start, for example, with a quantum-state electron in a 14 nm microtubule inside the neuron, or with some other quantum system protected, for example, by the Posner molecule we previously noted, and can then propagate to large-scale randomness in the brain. This would amount to having a truly random number generator in our brain.

This randomness has dramatic consequences. We have seen how life is most prosperous in the middle between randomness and predictability. This may also be viewed as the optimum between the needs for exploitation and exploration. If the brain is generally deterministic but quantum mechanics introduces some random noise, it will indeed push us toward the middle, namely to the regime of *unpredictable but not random*.

So, in fact, while quantum processes do not deliver the sought-after free will, or the metaphysical soul, they may just deliver enough potent randomness to save us from the death of being completely predictable by some AI in the near future (or for our brain to be directly proven to be a deterministic entity by some supercomputer predicting its internal processes in the far future).

This option of quantum randomness is obviously very different from the option of strong determinism on which we focused previously, as strong determinism has no randomness. It is also different from the possibility noted above of weak determinism, in which some real freedom may be assumed to be possible, as randomness is not freedom. It will be striking if future studies reveal that we are alive, in the sense of our new definition, due to the existence of a true random number generator in our head.

In the next quantum option, the survival of long-range quantum processes in the brain enables intricate possibilities such that extremely delicate quantum processes, for example, entanglement, may be at work in the brain. The existence of such a circuit in our brain requires a completely new understanding of the hindering (i.e., anti-quantum) process of decoherence and the ways evolution found to suppress it (e.g., as speculated by the new field of quantum biology). Such an option opens the door for a wide range of possibilities in terms of how the brain works. For example, entanglement would enable distant parts of the brain to affect each other, thus giving rise to a much more complex form of data processing and decision making, and perhaps it is even the key for consciousness. Whether or not such delicate quantum processes open the door for some form of freedom remains to be seen. At present there is no hint of that.

In any case, as I emphasized, it is clear that we still know only a small fraction of what there is to know, and of what we will know in the future. The study of quantum mechanics will surely bring an abundance of earthshaking new revelations.

Last but not least, and most speculative, the final option speaks of new physics beyond today's standard model. As I have discussed, this option may even give rise to the possibility that humans are living machines—machines because they are fundamentally deterministic, and living because some kind of new physics may perhaps not enable an external agent to predict the brain, as some parameters, for some reason, remain hidden or noncomputable, namely, with unknown values. The latter is of course an extreme case of new physics. While we clearly expect that new physics will be continuously uncovered by future generations, we have no way to estimate its features. However, whatever new physics is discovered, it seems that our two-layer model of the brain, in which one layer is responsible for unpredictability and the other for predictable determinism, will continue to be a good representation of the outputs of the brain.

Evidently, at this point in human scientific evolution, we cannot empirically differentiate between the above possibilities. However, the imminent near-future measurement of P will take us a long way in doing just that. For example, a high level of predictability in humans will teach us that noise and quantum randomness are weak in the brain. More importantly, an immediate consequence of a high level of predictability in humans is that they have no free will, or any other free agent within them, such as a free soul. These will of course be extraordinary revelations. If we also add to this some additional interpretation, such as our definition of *life*, the results of the experiment measuring P become omnipotent, affecting the very core of our species.

With what is currently scientifically known about the laws of nature and the structure of the brain, perhaps the best we can hope for in our quest not to be a predictable machine is that within behavioral constraints, established by our genetic makeup and our upbringing (our nature and nurture), giving rise to our personality traits, that within these boundary conditions of what our brain allows us to do, boundaries which we have coined filters (in our two-layer model), we have some seed of enabled unpredictability, brought about by some stochastic (random) process, whether it is classical (non-quantum) noise or inherent quantum randomness, which may also be termed noise. It will be quite astonishing, almost unspeakable, if we find that noise is a necessary condition that allows us to be human.

Perhaps we should again note that, as expanded on in Appendix B, scientists are working hard with probes such as EEG, MEG, and fMRI to quantify consciousness, the word we have been working hard to avoid because it is at present so vague. But even in the unlikely event that scientific work, in the foreseeable future, will bring the entire community of scholars to agree on what consciousness is and how to quantify it, will there be a consensus on how to utilize this internal mechanistic characterization of the brain as a criterion for human life? This internal characterization will most probably be much too complex for people to feel that it is trustworthy for such an all-encompassing definition. In the context of our new parameter P, an interesting question that should be asked is whether this probing of the internal mechanistic functionality of the brain will enable us to differentiate between brains that have been tested to have different values of P. If such a differentiation will not be possible for any functional definition of *consciousness*, then there are clearly grounds for suspicion that the definition found for consciousness gives only a partial description of the human experience. Specifically, in such a case, if we accept that *unpredictable but not random* is a necessary condition for human life, then such a quantification of consciousness will not be helpful in our quest to define different levels of human life.

In contrast, I believe that in the future a correlation will be found between P and the different states of the brain identified by the ever-growing body of empirical brain studies on consciousness. Such a discovery will teach us what in the brain is responsible for freedom or unpredictability. The latter gives rise to the possibility that a human may request that his or her brain be physically "fixed" in order to suppress P and ensure internal freedom, or at the very least unpredictability (e.g., by increasing the noise amplitude). For example, I expect that some external stimuli, or chemical, electrical, or even surgical intervention (e.g., an implant), would enable people to lessen their predictability. This would be considered some kind of medical treatment and may be termed the pill of the second kind, as it will come only in the farther future. This will only come after companies and medical providers understand that there is a huge market for such a pill, something that will probably happen only after humans can reliably measure their predictability and after they understand the horrific philosophical implications. It should also be noted that, unfortunately, dark agents may use the same knowledge to suppress freedom in the brain, namely, to increase P. In the next part of the book, we discuss what pill we may already take now to try and push us away from the state of a predictable machine. This may be termed a pill of the first kind, a pill that may be engineered by each one of us utilizing processes that are already at our disposal.

To end, I emphasize that even if one finds it hard to accept the proposed definition for *life* and *death*, the science of predictability still stands. One may choose his or her favorite interpretation of the experimental results, but if indeed, as I conjecture, a high level of predictability will be found in humans in the near future, the whole of humanity will have no choice but to consider the profound implications and consequences.

As the testing of predictability has already begun, by the business, political, security, and academic sectors, we may state that the most profound experiment ever conducted on the human species has been set in motion. It has the potential to give us some answers long before the complex study of consciousness comes up with clear-cut conclusions.

Once more reliable direct tests of predictability P are made available with the help of advanced AI and Big Data and we can measure predictability in a technologically mature manner, our understanding of life and who we are will be forever altered.

Part VI

CONSEQUENCES AND POSSIBLE PATHWAYS TO TRANSCENDENCE

9

The Pill

Dear Eve, as I noted previously, Confucius said that a human lives twice: the second life begins once we realize we only live once! But perhaps he was wrong, and in the future people will agree that there is also a third life, a life that begins once the human understands that our old common definition for death is a charade.

With the amazing rate of technological advance, it is only natural to ask what new medical constructions, that is, machines and procedures, will make possible for us in the very near future. Yes, longevity will skyrocket, and for those interested, changes to our external bodily appearance as well as performance-enhancing implants will become mundane. But could something more profound be expected? Can technology force us to redefine our own life and death?

Indeed, so far, we have surveyed a wide range of topics relating to the question of what human life is and is not, and found that a reformulation of our most cherished comprehensions is unavoidable.

In the introduction, we discussed the lack of accurate definitions for terms traditionally used to define human life, such as *consciousness*. It was therefore clear that we need to make a bold move, a move that would provide us with reasonably robust answers within our lifetime. In Chapter 1, we discussed how human predictability is an observable that enables us to make such a bold move. The philosophical and scientific basis for human predictability, which lies in determinism, was presented. Furthermore, the profound experiment now starting to take place to measure human predictability was described. In Chapter 2, we considered some of the common arguments against determinism and predictability. We found these arguments to be inconclusive at best. In Chapters 3 and 4, we took a philosophical pause to try to understand the limits of our ability to clearly analyze this question of human life, for example, due to personal biases and what we have been trained to consider to be common sense, and how we could free our mind by conducting a thought experiment. We drowned our brain inside a glass vessel to isolate the object of our examination, and with it our thoughts, from all external, irrelevant influences. In Chapter 5, we refined our question regarding the definition of human life and reviewed the present limits of scientific knowledge regarding death, namely, when human life stops. We found the present state of the relevant scientific knowledge to be quite rudimentary. In Chapter 6, we explored what is at stake and presented the possibility of a new definition for human life: human life is maximized when it is *unpredictable but not random*, that is, halfway between the extreme of randomness and the extreme of determinism, namely, in the middle between the roulette wheel and the predictable machine. Thus, a new definition for human life and death emerged based on a measurable (quantifiable) observable, predictability P, which AI and Big Data are already starting to measure. Such a well-defined, simple yet insightful, parameter enables us to avoid vague terms always called upon in order to define life, such as *self-awareness, consciousness, free will,* and *soul*, and move toward a more objective measurement and conclusion within our lifetime. This has extremely fundamental and pragmatic implications, and most importantly it allows us to engineer a new pill, distancing us from being predictable. In Chapter 7, we considered what may be the brain processes that could save us from the machine doom. These include neural network noise as well as the quantum mind with its inherent randomness, which may also be termed noise. Here, I also addressed your original question about the life of a single atom. The possibility of new physics and its effect on the brain were also briefly reviewed. In Chapter 8, a two-layer model of the brain was presented, which allows us to easily visualize the battle between randomness and predictability taking place in our brain. These are connected to the two fundamentally different operations in our brain, exploration and exploitation. Finally, in the conclusion section ending Part V, all the different arguments and findings were concisely listed. It will be quite astonishing, almost unspeakable, if we find that noise is a necessary condition that allows us to be human.

By the way, Eve, I frequently used the term *we*, as without your knowing it, you have become an active deliberating partner in this journey, roaming around inside my skull with your very insightful questions and directives. In fact, you have not only become a debate partner; you have also become an emancipator! You are probably laughing your head off thinking I am exaggerating, but it's true—you have forced me to fight my own chains and demons. As I noted in Chapter 3, the famous surrealistic painter René Magritte stated that he was trying to think as if no one had thought before him. With what concerns a new definition for human life, this is what we must do, and indeed what you have enabled me to do. I hope that this tiny confession of mine is not too awkward or embarrassing.

I believe it's now time to move on. Before you and I parted on the day you visited me, you added a very intricate request. It was a watershed moment for me, so at the risk of repeating myself, I echo my feelings regarding your request. I can still see your face shaded with bafflement, and I can still hear your voice saying the words quietly and slowly, as if you wanted to make sure I understood their importance. You asked that finally I should translate my scientific answer to the words of the humanities, explaining, "It is not because I don't appreciate or don't understand the words of the scientific language. It's just because I don't think it holds all the necessary wisdom required if the human race is to evolve successfully and prosper." I never told you this, but the more I thought about it, the more it sounded like an oxymoron, a paradox beyond my skills, a task that would demand of me, a physicist, to cut right through a Gordian knot.

I thought long and hard about your surprising request, and slowly I think I have managed to grasp what's behind it. Our history as a species is fascinating.[1] But what about the future? How do we strive to improve our species after we have solved the urgent perils of environmental well-being and the threat of weapons of mass destruction? Could our new observable, predictability P, have some impact on this future? In other words, if we are currently in danger of machine death, is there a way to heal ourselves? In the previous chapter

I proposed a two-layer model of the brain and suggested that in the future some external stimuli, be it chemical, electrical, or even surgical intervention, would allow people to lessen their predictability by increasing randomness, e.g., through enhanced neural noise. I noted that some scientists believe that noise in the brain is a precursor to "free behavior." Such a procedure would be considered some kind of medical treatment. This would only come after companies and medical providers understand that there is a huge market for the procedure, a realization that will sink in only after humans can reliably measure their predictability and after they understand the grave philosophical implications. I call this the pill of the second kind. But until that happens, could some sort of pill be engineered by humans for themselves with the means that are currently at their disposal? Let us call this the pill of the first kind. This is how I came to understand your rather surprising directive.

While very hard for me, the more I delved into your request, the more I found it to be a clever directive. Indeed, I recalled numerous thinkers who have noted that the language of the exact sciences, namely the language of logics, is not all encompassing. For example, the words of a London School of Economics professor, Ian Angell, who said that "the human brain created logic because it was useful, not because it was true. Logic is a product of intelligence, but the perverted causality of machine age thinking would have us believe that intelligence is the product of logic. Logic contrives to make simple and consistent what is not." Similarly, the great David Bohm, a theoretician of quantum mechanics, postulated that there was an "implicate order" hiding behind the "explicate order" that we observe and analyze, and that the latter is just a convenient representation of the former. I add here that the explicate order is the explicit order that is studied by the logical mind.*

If I take logic to be the main tool of the exact sciences, moving to the language of the humanities may award us with a more complete, perhaps more intelligent view. So I thank you for pushing me.

Consequently, I felt I had to understand more about the humanities and in what way their language and way of thinking are different from the scientific school of thought I belong to.

As noted in the beginning of this book, the philosopher Max Weber said that science can explain what *is* but not what *ought to be*. Perhaps this is also

---

*See, for example, https://en.wikipedia.org/wiki/Implicate_and_explicate_order.

what Bertrand Russell meant when he stated that everything has been figured out except how to live, and furthermore what Ludwig Wittgenstein meant when he stated that even after we solve all the scientific questions, we shall still be without answers to the important questions. Perhaps this is also why Plato and Descartes and Leibniz and Kant all pretty much agreed that science is secondary to philosophy. As noted in the introduction, I believe that what they meant to say is that the biggest drawback of rationality, which is the foundation of scientific and technological thought, is that it cannot help us define our ultimate goals but rather only the ways to achieve these goals.*

Not only philosophers thought that this gap, between what science can deliver and what we actually need, is of crucial importance. Artists, spiritual leaders, and novelists all understood that it is key to our future, and as such, perhaps key to truth. For example, C. P. Snow's *The Two Cultures* speaks exactly of this gap between the culture of science and technology and the culture of the humanities; an extremely literate and clever person in one of the cultures is typically illiterate in the other. One culture is usually concerned with measuring and understanding what *is*, typically in terms of the physical laws of nature and how to exploit them. The other is similarly concerned with what is, whether through history, philosophy, literature, or poetry (as well as the study of economics and politics), but is also concerned with what *ought to be*, typically in terms of the spiritual, ethical, and moral norms of humans and how they may be mapped onto human existence. The humanities approach the description of what is in a language that is also useful when trying to describe what ought to be, while the exact sciences do not. This seems to give rise to an existential gap that is creating havoc in the human species at the levels of individuals and societies. It is not by chance that Snow's book was listed as one of the one hundred books that most influenced Western public discourse since the Second World War; it seems people are indeed concerned that truth needs to be multidimensional and all encompassing, but not enough is done to bridge the gap.

Healing seems to require an attempt to bridge the above existential gap by fusing together the exact sciences, producing the technology that can make

---

*As an interesting anecdote, let me mention that the fact that the logical mind is not very good at defining ultimate goals was emphasized by the Dada movement in their protest against World War I. But one does not need to search for anecdotes: modern history is full of examples of very rational people, leaders as well as simple people, such as the employee of the large chemical company IG Farben mentioned in Chapter 5, who had misguided goals.

clear-cut measurements of the state of us humans, and the humanities, which could produce the crucial understanding of what ought to be. Phrased differently, by utilizing the powerful realm of the exact sciences, we can measure who or what we currently are and then make use of the realm of the humanities to understand how we can heal ourselves through a deeper understanding of what we ought to be.

But if you want to heal yourself, you first have to accurately define what is currently wrong and where you want to go. That is, you need to define exactly what *ought to be*. For example, a pill against headaches was developed because we could readily define that living without headaches is what ought to be. Indeed, when a disease is physical, it's easy to know what ought to be. But when all bodily functions are okay, like for zombies, it is much harder to define the disease, and consequently much harder to define what ought to be, where what ought to be is what we should be fighting for.

Therefore, the first thing to do before engineering the pill is to understand how sick we are. We need a quantifiable observable (i.e., one that can be measured by instruments to give a clear-cut number, a specific value) that can be a good empirical measure of the reality concerning the life of an individual, namely, is he or she a machine? In addition, at the same time, our quantifiable observable should allow us to define in numbers what we think ought to be. We can then compare where an individual stands relative to what they ought to be, and this provides us with a clear determination of where any individual lies on the new scale of life. This comparison is similar to our use of a thermometer: we know what temperature we have, and we know what temperature we ought to have, and consequently we know how sick we are. Similarly, a new technology measuring our new observable will tell us how far from being maximally alive we are.

Using predictability P as our observable, we may begin to bridge the gap between science, with its measurable observables, and the philosophy of what ought to be by defining P = 0.5 as what ought to be.* By extrapolating current scientific and technological advances to the point that they will be able to tell us much more about who we are, and even state how alive we are, we are able to make tremendous steps toward stating our goals with scientific accuracy and thus designing the pill. In other words, we can now discuss in more prac-

---

*Recall that in the context of our new definition for life, we found that P = 0.5 is a necessary condition for life but not a sufficient one.

tical terms what is required of us when we come to concoct a good recipe for a pill to fight predictability and the machine death that comes with it.

Human life must be the opposite of machine, and machine is predictability, so predictability is death. Observing people, it seems they are much closer to the P = 1 extreme of predictability than to the P = 0 extreme of complete randomness. If our goal is to live, then we must fight predictability. This is what the pill must do!

Thus, while the previous parts of the book were more tuned toward accurate definitions and considerations of a hypothesis in the realm of the exact sciences, this part utilizes the language and contemplations of the humanities, thereby attempting to complete the construction of the existential bridge noted previously, namely between the measured *is* and the hoped for *ought to be*, represented typically by the exact sciences and the humanities, respectively. It is this path that will allow us to design a miraculous pill that may hopefully heal us.

What ought to be was already defined as a number, an optimal value for predictability P (namely P = 0.5), and now we must ask, how can we portray what is and what ought to be in the language of the humanities? The humanities learn by carefully studying the most significant human attributes, so we need to ponder, for example, what can be learned about human predictability from some of the most extreme human traits, such as war, happiness, and so on. As predictability is the opposite of creativity, we can also start to wonder about creativity as a human trait that correlates with distancing ourselves from P = 1, and what the experiences in the human spheres are that enhance it, such as poetry and art and other relatively abstract activities. We discuss all this in the following.

Finally, Eve, I realize that my answer to you is rather long and perhaps too burdened with details. I apologize for that. The question you asked is probably the most important question one can ask, and I praise you for asking it, but it is also notably the hardest, so the answer is quite convoluted; there is no way around it, even with the bold move we have taken in bypassing popular yet ill-defined terms such as *consciousness*. If you are out of time, that's a pity, but I understand; nevertheless, please don't miss out on my final conclusions and personal note to you in the epilogue.

1. WAR

In the conclusion section ending Part V, I summarized what we know about the science of predictability. To start engineering a pill against predictability, namely, against the machine that is within us, I believe it is instructive to start by examining several cases known for their extreme human predictability, and I choose to start by discussing war. We should discuss war because, as I am sure you would agree, a typical soldier represents an extreme case of indoctrination (not to say brainwashing) in our civilization, and indoctrination is conditioning that is tantamount to predictability. In fact, the conditioning of a soldier so that he reacts exactly as expected (e.g., follow orders), namely so that he is predictable, is a precondition for an army to work. I held the rank of major in the army, and the latter preconditioning became crystal clear in terms of how I am conditioned by those above me and how I condition those below me.

I am not trying to promote a pacifistic agenda here. In fact, I do believe that some causes are worth fighting for, especially if it's in self-defense. I simply choose to look at soldiers as a case study in which predictability is groomed and rewarded as an example of predictability-enhancing processes in our society, which act in addition to any internal determinism brought about by the fundamental laws of nature as discussed in previous chapters.

In this section, as in the following ones, I will attempt to give some examples and considerations of how each of us can try and sense predictability in ourselves and in others, and how we might try to fight this measure of doom.

The question we should perhaps start with here is rather straightforward: How many of those soldiers who killed more than two hundred million people in the twentieth century alone had free will or, in our newfound language, did not have a high level of predictability P? Namely, according to our new definition of human life, how many of them were alive with a P value of around one-half?

But first to the facts. The number of two hundred million (and above) comes from the following breakdown of civilians killed: Soviet Union, 1917–1991, about sixty million; noncommunist China, 1928–1949, about ten million; communist China since 1949, about thirty-five million; Nazi Germany, 1933–1945, about twenty million; Japan, 1936–1946, about five million, most of them killed by their own governments. To that we should add some thirty-five million soldiers, and of course lots of other additions, such as from the Americas and Africa, so that in any case this is probably an underestimate.

I must say I could never understand how, around the world, children do not learn about this history of carnage. It is almost as if nations do not want to give the next generation an antidote against the conditioning. For example, how can it be that the many massacres and genocides the human species has perpetrated in modern times—like the Armenian, Greek, and Yazidi genocides, and those in Yugoslavia, Rwanda, Tibet, and Cambodia, and the extermination of the Romani people by the Nazis—are not taught? And what about the horrors the Belgians perpetrated in the Congo* and the concentration camps of the British in South Africa? Books like *The Forty Days of Musa Dagh*, describing the Armenian massacre, are for some reason not taught in schools. How can it be that no one so far has built anywhere in the world an independent, academically directed museum that would summarize all the massacres humanity has produced?† Perhaps one of the books on display there should be *Inferno* by Dante? And above the entrance perhaps we should put Hegel's words: "The only thing we learned from history is that we never learn from history." Perhaps additional quotes on the walls should be from thinkers such as Georges Bataille or Rosa Luxemburg, who predicted that the affluence created by modern society would be forever spent more on war than on things such as art or education. While governments are happy to build museums that describe atrocities committed against their own nation, as that fits the national narrative as well as the required indoctrination, it seems that no government, nor the international community as a whole, is in a hurry to describe human-induced carnage in general. Could it be that, knowingly or unknowingly, rulers and governments are not interested in invoking in individuals, and consequently in society, unrest, which in our language is tantamount to unpredictability? It stands to reason that extremely disturbing descriptions of the faulty traits of human civilization may indeed induce unpredictable thoughts, similar to the thoughts of the Dada movement in reaction to the First World War.

---

*They called the Belgian king Leopold the Builder, as he built many beautiful buildings in Brussels and throughout Belgium. Very few want to know where the money came from. It came at the end of the nineteenth century from the natural resources of Congo, for example, natural rubber. To produce the rubber, slaves were used. Millions died by murder or due to the conditions. A common practice in those days was to chop off hands of locals if they did not meet the daily quota. See, for example, *King Leopold's Ghost* by Adam Hochschild.

†As far as I know, there are two museums of tolerance, in Los Angeles and Jerusalem, the latter still having to prove it is capable of a universal message. There are plenty of museums about specific genocides, such as the numerous impressive museums in Berlin. There is also the US Holocaust Museum in Washington, DC, which stands apart as the United States was neither the perpetrator nor the victim. But there seem to be no museums telling the story of universal human carnage as a statement against the faulty human species.

It would be especially insightful to tell the story from the perspective of human conditioning or predictability versus independent thinking. We can start unraveling the story by asking, what was the first really big war where millions were slaughtered? It takes highly efficient soldiers to kill so many, and highly efficient soldiers are predictable. Perhaps Nebuchadnezzar, who came all the way from Babylon to Jerusalem to burn it, or was it in the conquests of Alexander the Great, or when in 31 BC the Romans defeated Mark Antony and Cleopatra who then committed suicide, or perhaps later on when the Roman Empire was established? Or perhaps the really big massacres started much later, for example, with Genghis Khan, who by some estimates killed forty to sixty million people (as an anecdote, let me mention that it is estimated by genetic research that he has some twenty million direct descendants). A similar number were probably killed by the Europeans in South, Central, and North America when they massacred the indigenous people, but this was over quite a long period. And what about 1487 in which it is estimated that the Aztecs sacrificed in one religious ceremony tens of thousands of people, where the total annual sacrifices probably reached 250,000 people? Perhaps we can start with Napoleon; his wars are estimated to have killed around five million people. We can decide to focus on the one that left in its wake the biggest monument: the 1813 battle of the nations, in which a consortium of armies tried to take on Napoleon after he left Moscow. Within a couple of days, one hundred thousand people died. If I am not mistaken, the monument that stands near Leipzig is as tall as the Statue of Liberty and was one of Hitler's favorite places for speeches.

Putting aside the story of Cain and Abel, I believe the first archaeological forensic evidence for murder dates back some half a million years. What was the reason? What went through their head? I guess it started with the mind of the apes, and since then our intelligent mind just got better at it. Was free will involved? Could our new technology for anticipating the human have predicted the action?

The conditioning of soldiers to be predictable means no skepticism. It is thus no wonder that, for example, American soldiers are typically not made aware of the speech made by American president Dwight Eisenhower on January 17, 1961, where he warned of the growing influence of the military industry in Washington: "In the councils of government, we must guard against the acquisition of unwarranted influence, whether sought or unsought, by the military-industrial complex. The potential for the disastrous rise of misplaced power exists and will persist. We must never let the weight of this combination endanger our liberties or democratic processes." Or, similarly, the speeches of Martin Luther King Jr., in 1965–1967, against the Vietnam War, where he said, for example: "The bombs in Vietnam explode at home—they destroy the dream and possibility for a decent America." Would Colin Powell have made his 2003 address to the United Nations on Iraqi weapons of mass destruction if he was not conditioned to be less skeptical? I do not believe he lied. Perhaps he was conditioned to the point of predictability.

As a notable exception, let us mention that the extreme situation of war also shows us from time to time that there are outliers, people who did not behave as expected. A famous example is that of Hugh Thompson and his helicopter crew, who noticed an ongoing massacre in which US Army soldiers tortured and killed several hundred unarmed Vietnamese civilians in the village of My Lai. At risk to his own life, he stood before the army soldiers in order to protect the civilians. Many years later, his actions were recognized and decorated. For obvious reasons, most outliers are never recognized and decorated. A well-known example is that of Israeli brigade commander Eli Geva, who during the siege of Beirut in 1982 refused a direct order to lead his forces into the city, knowing that many civilians would be killed. He was relieved of his command and was not allowed to ever return to army service in any capacity. A year later, Stanislav Petrov decided, against Soviet military protocol, to disregard the Soviet early-warning system, which falsely indicated that a US nuclear missile strike against the USSR was underway. He had a civilian background, and he later stated that his colleagues were all professional soldiers with purely military training and (in contrast to him) would have reported a missile strike if they had been on his shift. He was reassigned to a less sensitive post and suffered a nervous breakdown.

Would AI and Big Data be able in the near future to locate these outliers ahead of time? I am sure they will. China is already making headway in finding future political dissidents before they actually do anything meaningful.*

---

*To get a glimpse of the future pyramid of predictability, one may read a recent article in the *New York Times* about how China is already predicting the future behavior of outliers: "How China Is

Armies will surely want to strengthen their functionality in battle by reducing the variance in soldier behavior, especially when following direct orders. Taking away the outliers, three of whom I presented above, is surely bad news, as these outliers keep armies, at least to some minimal extent, moral and truthful, and, more than these armies would care to admit, these outliers even improve their performance.

Predictability in war is of course intimately related to the issue of obedience. Predictability of violent acts is not just about the preconditioning of loyalty, as described above, but also about the knowledge of what types of social pressure or real-time manipulation can control humans, giving rise to predictable violent outcomes in real time. The famous Milgram experiments of the 1960s, in which people were convinced to administer electric shocks to other people, deal with obedience but in fact could not have given the observed results without humans being predictable. Since then, numerous experiments of this type have shown that a high degree of extreme manipulation of people is possible. For example, in 2016, in the TV program *Pushed to the Edge*, people were persuaded to push other people off a rooftop to their death. This is in fact quite similar to our description in Part I of the work of Edward Bernays who took Freud's ideas and successfully manipulated public opinion on numerous matters. Mastering such influence over people can only originate from the knowledge that situation A would cause outcome B. This is the definition of predictability.

To conclude, even if there is no universal determinism in our brain causing everything since the Big Bang to be predetermined, it seems that our life is full of predictability enhancers. For example, every pyramid structure of an organization, whether a Wall Street financial firm, a company that makes cars,

---

Policing the Future," *New York Times*, June 25, 2022. In the introduction, I noted that Eve's avatar suggested that, since predictability is such a powerful tool, eventually everywhere, not only in China, there will be a social pyramid of predictability, where each layer of society can predict the layer beneath it, as each layer has more harvesting and predicting technology than the one below it.

or an enterprise like NASA, requires predictability of the base for its stability and feasibility of governance. Would the *Challenger* disaster have happened if there was less predictability among the employees cutting vertically through the hierarchy? We can only speculate as to the answer.*

Civilization itself is based on codes of behavior, which are tantamount to predictability. The flag and anthem are also intended to increase loyalty, which amounts to predictability. As we noted, the military is an extreme example of a pyramid structure that requires predictability in order to function. The conclusion is that, even without universal physical determinism represented by the Laplace demon, our brain is conditioned for predictability from a young age, by cultural, religious, national, business, political, and, as discussed above, military constructions. Perhaps even by love and hate.

If we agree that predictability means machine, and machine is the opposite of human life, then we, the human species, will have to try and go beyond our previous path. If our predictability is not due to a strong universal determinism, the door for such a change may be open. In fact, Jean-Jacques Rousseau said something quite similar, noting that the same line of thought that frees humans from the determinism of an a priori fixed nature, which is the same line of thought that turns humans into victims of society, is in fact the line of thought that hints at the option of liberation, the possibility of emancipation.

We may add here that speaking of emancipation is essentially tantamount to speaking of the pill, the anti-predictability pill we have been dreaming of. This is so because when Rousseau speaks of victims of society, he means that society has forced its people to abide by its rules, and this means they are predictable.

Furthermore, we may note that *emancipation*, like *consciousness* or *self-awareness*, is also a vague word, as what is enabled by it was never clearly defined. Perhaps free will? But one thing is clear, our measurable observable, predictability, if it reaches a value close to $P = 1$, is the opposite of all of them. It is thus able to clearly tell us who we are in the most candid, unambiguous, biting manner.

---

*Remember the story of Nobel laureate Richard Feynman, who was rejected by the army and eventually not recruited on the grounds that he failed his psychological examination (some claim he faked his mental instability). Many years later he discovered why the *Challenger* exploded. One may wonder if Feynman's "lack of normality" is not tantamount to "lack of predictability," which enabled him to make this and other discoveries.

2. HAPPINESS

In the introduction, we noted in passing that a predictable, thus mentally dead, person could be a happy person. However, as happiness is such a strong emotion in the human sphere of existence, it deserves a much closer look. Previously, we talked about war as an extreme situation in which our human actors are predictable. This shows that the potential for predictability in humans is high. Even more than a soldier of war, a strong initiator of, or sufficient condition for, predictability is happiness, for when we are happy, we do not seek change, and change is dynamic and hard to predict, while lack of change gives rise to the exact opposite. I believe it is not by chance that many innovators, whether artists or technologists, had miserable childhoods or adult lives. But different from war, happiness could, as we shall see, also be used to fight predictability. In this sense, happiness, if used correctly, could become the first ingredient in the pill of the first kind which we would like to engineer.

Examining happiness in an objective, not to say critical, manner is tricky, as we have always aspired to achieve happiness. It has become almost a holy word, a cultural axiom, a meaning-of-life postulate.

There are even philosophies, such as utilitarianism, intended to maximize happiness for the largest number of people (as opposed to hedonism), stating that maximizing happiness is the only criterion for a successful action.

But what is the definition of this happiness? It is as vague as speaking of the emancipation of the mind, self-awareness, or consciousness. We humans have always set our heart on obtaining it, although we as a species and as individuals have not done enough to define it and consider its consequences.

Not so long ago, when people were restless, doctors had these very aggressive psychiatric tools, like electric shocks that sometimes actually erase your brain (or at least its memories), or even disconnecting parts of your brain in lobotomy procedures. Now they prefer to use drugs. In fact, the way things are going, in the not so distant future, they will be able to control the internal drug delivery in the brain. For example, some doctors have started treating depression with deep brain stimulation by tiny implanted electrodes, where the goal is to return the brain's metabolism back to a state of balance. Humans have always searched for ways to have everyone at ease. Isn't this exactly how people behave when they are content, and isn't contentment simply some inner chemical balance that we may even refer to as optimal internal drug delivery? And frankly speaking, isn't this feeling of being content what we mean when we say *happiness*? As I noted above, perhaps it is not by chance that most if not all of the great artists did their best work when they were in some state of lack of happiness.

I recall that a professor of poetry and literature once told me that happiness is a statue. It was clear that this was her way of cursing happiness, but I did not really understand back then why she was cursing happiness, as she seemed, after all, to be a happy woman. It took me a long time to realize that she may have been cursing happiness for two reasons: first, personally, because it was blocking her, and second, in the name of humanity, as humanity needs unblocked people to advance.

Let me continue to try to see through the ideal of happiness. In fact, it could be that the great hour of happiness has not arrived yet. As humans become content with material things, and as material things become easier and easier to supply, it stands to reason that it will become increasingly feasible to make humans content on a massive scale. With the advent of extremely high computing power, this supply of things we want can be adjusted to meet the personal needs of each of us, thereby further increasing our level of satisfaction. But this is not limited to just material things. As we discussed, the enormous computing power of AI with Big Data will enable it to follow each of us as we grow up, to understand our needs through our activity, and to supply those needs, for example in terms of occupation or even a compatible partner. In fact, Big Data and AI will be so powerful that we may safely conjecture that they will also be able to understand and supply our mental, spiritual, and psychological needs.

Happiness will be widespread. As long as we avoid environmental catastrophes or the elimination of ourselves with some killer virus we invented or in a nuclear holocaust, it stands to reason that happiness will be widespread. It may take a hundred years or a few hundred years, but it's coming: the dictatorship of all times, the absolute dictatorship of predictability. I am not talking about a political dictatorship, where some ruler rules over the masses because he is able to supply happiness, and they are thankful and repay him with their loyalty. No, I am speaking of an internal dictatorship—which perhaps then allows the political dictatorship—an internal dictatorship of happiness that does not allow the mind to be skeptical, suspicious, curious, or creative, to wander off seeking new ideas, changed concepts, and aspiring to achieve a different, higher form of life. If everyone is happy, no one will ever want change, and without change, things are much more predictable. Consequently, according to our new definition of human life, the dictatorship of happiness will kill us all.

We have already discussed the work of Thorndike, who showed that responses that produce a satisfying effect in a particular situation become more likely to occur again in that situation. Similarly, a scientist named James Olds took rats and connected an electrode to a part of their brain where the electric stimulation was rewarding for them. They taught the rats to push a switch every time they wanted the stimulation. The rats ended up preferring the stimulation even at the expense of food and water, even after they were starved. Sometimes they would press the switch until they collapsed. If we are happy, we want nothing else but to keep doing whatever makes us happy.

We could thus adopt a definition or a measure of happiness which is that we stop having *libertas disputandi* inside our brain, namely, the freedom or the mere wish to argue, to have opposing opinions, with some of those opinions asking for change.

To assist in the discussion on happiness and its correlation with predictability, I would like to briefly examine aesthetics, as the latter is similar to happiness in the sense that it aims at some pleasing chemical balance in the brain,[2] while it seems to have an easier definition.

So what is aesthetics, and what is art, and what is their relation to happiness?

Obviously, hundreds and thousands of books and articles have been written on the topic of the philosophy of aesthetics and of art. I do not pretend to know them all. In the spirit of this book, and while giving due respect to the brilliant people who have worked on this topic, let me try to be succinct and focus with commonplace, nonacademic language on the relation of aesthetics and art to the dictatorship of happiness and predictability.

The word *aesthetics* probably dates back to ancient Greece where it meant something like the perception of the senses. Philosophers like David Hume and Immanuel Kant made the connection between the sensation or the ability to detect the ingredients in a composition of some object (this could be visual, audio, text, or action) and the ability to feel pleasure because of this object. It is this aspect of aesthetics (i.e., pleasure) that I am about to focus on, as pleasure is a big step toward happiness.*

Aesthetics of form, of shape, of a moment, of an experience, provides us with this sought-after chemical balance when it fits some internal model that we have adopted through genetics and by the molding of the environment within which we grew. "Adopted" is perhaps not the appropriate word here, as it implies intent, while this internal model of what is pleasing to us was imprinted into us and is almost beyond our control.

I am speaking of the standard form of its semantic meaning, an aesthetics that gives us instant pleasure, a fast-food restaurant for chemical balance. We may even call it idle or lazy aesthetics, because it takes the easiest route to influence a person. Its goal is to be pleasing to the mind. As with kitsch (see Appendix D), the reaction is easily predictable once you know the specific inner model of an individual, which we sometimes call taste. The point is that the reaction to aesthetics, in its idle form, is simply satisfaction, and it in no way induces things like skepticism and searching.

As with happiness, dissecting aesthetics is quite hard for the human mind, for aesthetics is beauty, and how can one speak against beauty? Speaking of aesthetics, we should probably speak also of art. What is art, and what is it for? I shall again try to squeeze this extremely complicated question, which has been dealt with by brilliant thinkers for generations, into a few simple and practical thoughts relevant to the matter at hand.

Evidently, artists make art for a very wide range of reasons.[3] Let me try and summarize some of them. Perhaps art arises from some personal need of the artist, some inward reflection relative to which the interaction with the audience observing his art is of little importance.† Obviously there is some strong internal drive in any artist, sometimes called "compulsion to create," that is

---

*The fact that pleasure is chemical, and that some chemical state is responsible for inducing it, is becoming more and more scientifically understood. Even in the conclusions of the *Stanford Encyclopedia of Philosophy*'s entry on pleasure, one may find the following: "The prospects seem good for new and deep scientific understanding of pleasure and of how it is organized in the brain."

†In the early nineteenth century, there was a famous French slogan, *l'art pour l'art*, namely, art for art's sake, with the idea that art should be divorced from any moral, political, didactic, or utilitarian purpose.

independent of the outside world, but I would guess that many if not most artists would prefer to seek higher ground. If we talk of the audience, perhaps the motivation is recognition and honor, similar to what, at least in part, pushes forward the scientific community. I would again speculate that for many if not most artists, this is not the main thing that drives them. Furthermore, in my humble attempts to reach the pure core, I will put aside any need for some sort of immortality. What then? What about art as a pleasing medium intended simply to provide joy? Again, if I had to hazard a guess, I would say that most artists would not agree. Assuming that artists are genuinely aware of what drives them, what would they say? Perhaps sending a message, inducing some change, making people think, making people feel, making people better? Isn't that why conceptual art and pop art came to life? Perhaps some artists would even state that they are working for good?* I believe most artists would prefer these last motivations over all others.

Let's assume that sending a message, working for good, is at the core of the arts. But how do you deliver this message? Do you aim for pleasure, for that chemical balance? Or perhaps you need to be disruptive, to sting, even insult, to break that very same balance? Even if we believe that eventually the only sustainable situation is that of chemical balance, such disruption of the old chemical balance could excite the brain to a level where it may find new points of balance, just like mathematical equations may have several solutions. It is similar to a ball placed on a surface with numerous valleys and hills (like in an egg carton). It is stuck motionless in one valley or hole, but if we give it enough energy, it may roll over a mountain and reach a new valley where it may reside. The point is that being able to move to these new points of chemical balance will suppress predictability, as it is hard to predict to which valley of chemical balance you will move. Hence, multiple valleys are like some glitch in the deterministic software.

I am thinking of my field of experimental physics, and indeed I always believed that each challenge has numerous solutions. In fact, when I think of

---

*I must note that I use the word *good* cautiously, as philosophers have already suggested that no action is truly driven by good and that only goodwill itself, namely the thought itself, may be considered as representing pure good.

some solution, even if I really like it, I always try to leave the table, leave the office, and force myself to come back the next day with an alternative solution, and then I let them fight each other.

So every valley, every such point of chemical balance, describes a point in which the brain is happy, but some of those valleys may be very different and may, for example, include skepticism as a condition for happiness.

Those who create aesthetics, the manufacturers of pleasing calm, will say that to create a habit of beauty is to fight the ugly by way of internalizing norms, by way of effective conditioning. It is conditioning because it is a form of education that does not base itself on options, skepticism, thought, or complex circles of for and against. It is a simplistic form that does not connect itself to questions regarding the meaning of existence or the lack thereof. It thus delivers a predictable outcome, but if this predictable outcome is to hate the ugly, that, they argue, is a worthy path that should be pursued. These proponents of aesthetics would state that this is indeed the right way to fight ugliness, for what good did complex independent thinking, which is tantamount to unpredictability, do for humankind? They would argue that the things that require fixing have been identified ages ago, and still folly and barbarism are everywhere, and with them ugliness, and to be successful in fighting ugliness, we must make people more predictable.

Indeed, I think that the answer to the question of whether to adopt the simple conditioning of aesthetics or the much vaguer disruptive conditioning as a path for a brighter future previously depended mainly on the level of trust one puts in humans. But now that we understand that predictability is our enemy, and since we understand that aesthetics—which manufactures the chemical balance of pleasing calm—is synonymous in this sense to happiness, and they are both precursors if not initiators of predictability, we have another reason not to adopt the simple form of aesthetics and happiness.

As predictability is machine and machine is death, we should become proponents of those who rebel against any form of pleasure and happiness that does not inherently include revisiting, reexamining, and questioning norms and truths. We should thus only stand with those who rise up against addiction to the drug of simple happiness.

So what is this alternative aesthetics and art that bites and stings and allows the ball to go to the next valley, thus giving rise to unpredictability? To put a toilet in the center of an exhibition, a trend started by Marcel Duchamp, so that humans are confused? Edvard Munch's scream so that humans see the abyss? Abstract and surrealistic forms, like Magritte's man with an apple head, or a castle floating in the air, so that humans feel lost without their logical base? As the human changes, what bites and stings will change, but it will always have to go as far as possible from what constitutes kitsch for that generation (see Appendix D).

Finally, pleasing calm that does not enable excitation can also come from a predictable environment that creates mental comfort zones. Even in architecture you find visionaries such as Friedensreich Hundertwasser, who claimed that the straight lines of the standard architecture of human dwellings suppress the human spirit. He actually used the word "subdues." He stated that such efficient forms of architecture are more fit for machines.

Evolutionary dynamics, which may be described as biological nature searching for the next-possible, in fact portrays a very similar narrative to our array of possible valleys of chemical balance in the brain. Stuart Kauffman called it the adjacent-possible, and, interestingly, he said that the next-possible would always be more complex. The similarity is, however, not complete. For example, in typical biological systems, random changes and adaptation to the environment push the system to the next valley, while in the human mind this seems much harder. Something seems to be blocking mental excitations, and this brings about predictability. These blockages may simply be described as the filters in our two-layer model of the brain (see Section 8.1).

Perhaps this blocking of mental excitations comes about because we are not built to survive these complicated models of happiness suggested above. Perhaps nature already tried. Or perhaps the required excitation is just too big, too unstable. Perhaps disruptive aesthetics and art are not strong enough to create such excitations. I would not like to believe, as many creative people do, that mind-altering drugs are the only way to wander off to new valleys (see references in the next section).

If there is no universal determinism jailing us in some dungeon, and if at the same time we are doomed to search for that pleasing chemical balance, we may still be able to fight predictability. Let the artists of the future be built from orthogonal models, completely without overlap to those of their audience. Let the space of the exhibitions be filled with provocations, disturbances, screams of the mind.

Similarly, let the happiness, which we will always seek, include a never-ending desire for skepticism. Let it follow Descartes's statement, "I doubt, therefore I am." If through education and the emotional environment of a child we could build a new model for happiness that requires skepticism and divergent thought (creativity) in order to arrive at happiness, we may find efficient ways to combat predictability without mind-altering drugs.

In fact, long ago, people already suspected that there are several fundamentally different types of happiness. For example, while hedonistic happiness has to do with pleasure (i.e., enjoyment of experiences), eudaimonic happiness comes from finding purpose and meaning in life, that is, from fulfilling our potential as humans.

We are in a race against time. Before the program to produce simple (hedonistic) happiness is completed with the help of AI and Big Data, by those who want to be in control or simply to profit, we need to produce a complex egoism whereby the needs of happiness include critical thinking and opposition, and the wish for change.* This is the only way that the tidal wave called happiness can be turned to work on our behalf, to become an ingredient of the pill.

3. CREATIVITY

We continue with our effort to design a pill of the first kind, a pill with which we can at present fight predictability and the machine death that comes with it, with the meager means already at our disposal. In fact, humans have been thinking of the problem of predictability, but in different terms. Instead of being unpredictable, they talk of being creative, but it is quite clear that the two are nearly synonymous, so much so that our measurement of P may in fact also be viewed as a measurement of creativity. People have also used

---

*And, of course, no text on happiness can avoid mentioning the seminal work by Aldous Huxley, *Brave New World*.

the term *divergent thinking*, or simply put, thinking outside the box. This is exactly what is meant by the excitation to additional valleys in the previous section. If we are controlled by a mainly deterministic brain, creativity may not change our eventual predictability score (which we termed p, see definitions in Section 1.2), but it may lower P, as the volume of information we need to collect on an individual before we can make high-quality predictions concerning his or her future preferences will be much higher for a creative mind. Hence, for all practical purposes, the individual will be less predictable to his peers or to AI.

Mind-altering drugs have been used by innovators in art and technology as a way of obtaining creativity (see Section 4 of the introduction).* They may certainly be abusive to our body, but they nevertheless show us the amazing potential of fighting against P = 1. Beyond the redefinition of *happiness* as a tool against predictability, as discussed above, in the following I suggest additional ways to fight predictability, although it is clear that much more work needs to be done in order to understand the best pathways. I believe that those who wish to be unpredictable without using drugs should invest time and effort in exploring their options for the pill. Specifically, regarding creativity, as over the years numerous techniques have been developed to enhance it, these techniques may also be thought of as techniques to fight against P=1.

I was once asked to give a course on creativity at the Bezalel Academy of Arts and Design in Jerusalem, and I told students that although there is no formula or prescription, the best method is probably to master the art of opposition, opposition to others but mainly to yourself. I told them to map the space of possibilities concerning each decision they need to make (*parameter space* in the language of physicists) and choose the road not taken (in the words of Robert Frost). In short, to revolt, to excite themselves to that next valley of balance, to climb the mountain to that next-possible.

As examples, I would like to tell you of two projects the students submitted based on the above. The first student invited us all for breakfast and made us

---

*Two examples are Pablo Picasso, who smoked opium and stated that the smell of the thick opium smoke was "the most intelligent smell in the world," and the famous surrealist artist Salvador Dali, who said, "I don't do drugs. I am drugs." On the science behind mind-altering drugs, see note 3 in the introduction.

promise to come hungry. When we came to class, there was plenty of great food on the table. However, she quickly locked all our elbows with cardboard paper tightened by tape, as if our hands were kept straight by plaster. She said she was revolting against elbow movement. We were hungry, and we spontaneously divided into pairs, each person feeding his or her partner. The second student asked us to come in the evening for a disco party. He put in place fancy lighting and a great sound system, but as we were about to dance in pairs, he made us wear goggles. His goggles were not virtual reality or augmented reality goggles but periscopic goggles, whereby in front of each eye he placed two forty-five-degree mirrors so that you could only see what was behind you, with two outer mirrors placed such that the rear view was not blocked by your head (such forty-five-degree mirrors are reminiscent of the arrangement in reflex film cameras at the end of the twentieth century, or of a submarine's periscope). After some confusion, all pairs of students started to dance back-to-back, as this was the only way they could see each other. He said he was revolting against face-to-face dancing.

Psychologists, scientists, and, more recently, the high-tech and business sectors have long been on the path of trying to understand creativity and how to enhance it.[4] The constraint-imposing filters, which we have been hypothesizing in our two-layer model of the brain, have been given all kinds of names, such as *cognitive fixedness*, which includes functional, symmetric, and structural fixedness. These have to do with the filters that affect the way we interpret reality when we try to solve a problem. Numerous protocols such as systematic inventive thinking (SIT) have been designed to force the filters down. It is interesting to note that as businesses scramble for more creativity, and as creativity is a strong suppressor of predictability, it may just be that the business sector will be inclined to view our new definition for human life favorably. In fact, some articles already talk of enhancing serendipity as a way to enhance creativity, and this of course is very similar to our analysis concerning the importance of randomness.* Obviously, the industrial sector will have to realize that

---

*It is interesting to note that some books and articles, even in financial journals, have suggested that more randomness, in the form of serendipity, should be engineered into the life of people (e.g., in the business sector) in order to foster creativity. See, for example, http://www.economist.com/node/16638391, https://www.economist.com/books-and-arts/2017/03/09/in-praise-of-serendipity. External serendipity may assist internal randomness, and this in turn fights against $P = 1$.

creativity cannot be enslaved to very narrow goals, and if you want to truly enhance creativity, you cannot imprison the mind within the confines of simple Excel sheets of timelines and profit expectations. This will just move the mind from the exploring mode to the exploiting mode, which, as discussed, are connected to the two opposite extremes of the spectrum of predictability.

Finally, let us recall that creativity is associated by some scientists with internal random (spontaneous) fluctuations in the brain (see Section 7.1 on noise). It remains to be seen if those could somehow be enhanced. Indeed, as noted, in the future, as part of the pill of the second kind, more potent interventions such as brain implants would enable us to increase our natural brain noise. However, for now, within the realm of the pill of the first kind, we can train ourselves to reduce the filters in our two-layer brain model. This is the business of revolting, which I tried to teach my students. Indeed, in the papers cited in the section on noise, the authors also hypothesize that these spontaneous fluctuations must work within boundaries and constraints determined by personality traits and other predetermined features of the brain. If some of these constraints can be changed by training our mind, the door is open for increased creativity.

The bottom line is that with creativity we are in the same situation as with Pascal's wager, which was mentioned earlier. We don't know if we have some freedom or ability to be unpredictable, but in a scenario of uncertainty, we should assume we do. If we don't believe there is a possibility for $P = 0.5$ and it does exist, our losses are immeasurable.

4. LOGIC

These instruments of the near future that can measure P are not yet available, at least not officially, at least not to the general public. As conjectured in this book, they will be made available soon. Yet there may already be some poor person's test that we could apply to ourselves. And if we find that we are not fully alive, is there some technique for resurrection, namely, a pill for lowering our predictability?

We have already talked about redefining happiness, about excitations to the next valley of chemical balance, and about creativity and divergent thought. Is there more we can do?

I would say that there are many forces that may work against predictability, even rage, as in the words of Dylan Thomas. But the strongest adversary to predictability P is most probably passion for imagination, as imagination requires a random seed. So increased imagination forces the brain to enhance and utilize its random core. This forces the brain to lower its filters.

Einstein said that imagination is more important than knowledge. What limits our imagination? I think it is first and foremost an all-encompassing belief in the logical mind. This limits our ability for imagination as the extreme logical mind will only consider what it deems to be currently possible. In fact, logic hurts us twice. First, it increases predictability simply because an external agent can identify what the next logical step is, while it is hard for an external agent to predict a less logical thought. And second, because, as stated above, even within our mind, logic seems to suppress the ability for imagination. It is almost a zero-sum game.

We are all zealots of logic, and for a good reason. The logical mind has indeed brought us far. To begin with, it has allowed us to survive for millions of years, for how can you hunt without a logical mind? More recently, it has emancipated us from superstitions and fear of the unknown, and thus we owe it a great debt. It could very well be that we are not just consciously admirers of the logical mind; it could be that evolution enhanced in us the logical mind in order to survive, so that it is deeply entrenched within us.* But could it perhaps also hold us back if we do not strike the right balance between it and other pathways of thought?† Or, in the language of our two-layer model of the brain, could the filters based on logic and tuned for exploiting be too strong, consequently preventing us from exploring, which requires imagination and a seed of randomness?

Einstein also said, "Invention is not the product of logical thought, even though the final product is tied to a logical structure." In fact, what he was actually saying is that invention, which comes from exploring, and the logical mind are two different things.

---

*Obviously there is also a significant irrational component to us, but it seems we are constantly in the process of fighting it.

†There is some evidence that the left part of the brain is more tuned to analytical (logical) analysis, while the right part is more intuitive, seeking an understanding through insight. In this language, the improved balance we are seeking is between the two brain hemispheres.

Similarly, Leibniz, one of the great thinkers of our species, thought that the "use of reason" is only part of what he called the art of thinking.

This is most probably true, as there are evidently many things that the logical method cannot comprehend simply because humans have limited analytical capacity, not to mention limited sensory capabilities, while reality is infinitely complex. But even if we had infinite capacities, I would still believe this to be true.

Indeed, perhaps logic is not only limited by practical considerations. Perhaps there is some deep fundamental limit. A suspicion builds inside of me that it might be that this is the reason that in order to secure the survival of life, nature opted for a strategy of mistakes or random mutations, as any mechanism of adaptation based on some logical defense against threats is bound to eventually fail against the infinite variety of future challenges. Human imagination and associative thinking may be the analogue of nature's strategy.

So, in what ways may we distance ourselves from our complete enslavement to logical thought and attempt to reach a more advanced balance that would be less predictable? Let me speculate that there are quite a few paths available to us, and furthermore, let me offer a rather radical example and hypothesize that poetry is such a path. I am of course speaking of high forms of poetry that are very much distant from kitsch. Indeed, I believe that by admiring poetry to the extreme, this could be achieved. This means that I would not be surprised if we eventually find that high forms of poetry to some extent have the same effect on the brain as mind-altering drugs (assuming, as many artists do, that indeed mind-altering drugs improve creativity by suppressing the logical mind). This will have to be proven by machines such as fMRI and EEG. Currently, such studies have already begun to show that poetry excites the right part of the brain.[5] I also hope scientists will look at whether a correlation or perhaps even a causal relationship exists between high forms of poetry and the strength of those internal brain fluctuations that we discussed in Section 7.1 on noise.

Poetry may serve many purposes beyond the mere joy and satisfaction of the poets themselves. Poetry can portray the beauty of humans as well as their ugliness, it can tell us of the wonders of nature and the universe, and it can reveal the mesmerizing lure of love. But I believe it could also lower the predictability of the mind.

The fact that poetry is so important to our survival has been pointed out endless times, but perhaps this is the first time where a quantifiable reason is given, namely, predictability.

I recall that Magritte, the surrealist painter, said that in the hierarchy of truth, poetry is above physics. If I had to hazard a guess, I would say that, like with the Eastern interpretation of spiritual enlightenment, he thought that the fact that poetry is only half logical makes it a more encompassing truth.

Wanting to describe poetry, I shall first try to do it justice by quoting a few words from the speech given by A. C. Bradley, an Oxford scholar, in 1909, titled "Poetry for Poetry's Sake": "For its nature is to be not a part, nor yet a copy, of the real world (as we commonly understand that phrase), but to be a world by itself, independent, complete, autonomous; and to possess it fully you must enter that world, conform to its laws, and ignore for the time the beliefs, aims, and particular conditions which belong to you in the other world of reality. . . . It is a spirit. It comes we know not whence. It will not speak at our bidding, nor answer in our language. It is not our servant; it is our master."

Going back to our issue of predictability, the most beautiful definition for high poetry, according to my subjective viewpoint of course, is that it is a text on the borderline of connectivity, namely, of what may be put together by the human mind in a logical manner, connectivity on the edge of possible deduction. Abstract poetry is the extreme example of that. Haiku poems are sometimes like that.*

I think high poetry aspires to go beyond the edge of the possible, as the possible is what was, not what will be. It is the hypothesis that the old medium of delivering a thought or an idea (i.e., through some logical stream of bits), either inside of us or outside, is maimed.

One may go so far as to say that high poetry is an attempt to disqualify the logical mind. If we allow ourselves to use a vague concept, poetry is an attempt at spiritual enlightenment. Similar to an abstract or surrealist painting that tries to steer us away from any logical interpretation, poetry is a mind-altering drug. I believe that those future instruments we have been discussing will show that poetry suppresses predictability.

---

*Haiku poems are a Japanese tradition. For Haiku-style poems written by physicists, see Christopher Cokinos, "Poetry in the Abstract," *American Scholar*, April 15, 2021. I recommend that everyone give it a try. Here is my humble attempt: The river knows its course. Time knows not of fear. Beautiful is the wind.

I once heard someone say words of enchantment regarding poetry, and unfortunately I do not remember who said them (and Google seems not to remember either): "It's ridiculous to think that poetry can change the world, but it's just as ridiculous to think that the world can change without poetry."

I believe the above discussion regarding high poetry also applies to any low-connectivity, alogical art, such as modern dancing, abstract or surrealist painting, and so on. Foolishness, like walking down the street with a cooking pot on one's head, may also contribute, or writing an ode, a lover's text, to some inanimate object such as the floor tile you are now standing on, or the zipper you are now using.

To conclude, imagination is important for suppressing predictability P, and the logical mind seems to weaken imagination. It may even be imagination-phobic. As the logical mind has been so successful, it is not easy to convince our mind to restrain this omnipotent part of our thought process until a balance is reached. But if we are to fight predictability, we must.

5. SOCIETY

Finally, while we have been focusing throughout the book on the individual and his or her life, it is clear that the future less-predictable human we have been talking about will also have a profound effect on society. This human will not only make life for totalitarian regimes much harder, but he or she will also help improve and push forward nontotalitarian regimes. Those who really push civilization forward are typically individuals on the fringe who are born from diversity, not from the mainstream, which exemplifies homogeneity and predictability.

The same question we asked throughout this book concerning the predictability of the individual may now be asked concerning society. Namely, how predictable is society? Typically, the spectrum of human behavior on any topic may be represented by a normal distribution, sometimes called a bell curve, and may be approximated by a Gaussian function, which looks like a mountain. It is usually plotted on one of those standard x-y graphs, where the x-axis gives the value of some human feature, trait, or decision (e.g., love of trees from one to ten or number of hours in front of the TV), and the y-axis presents the frequency with which a specific value appears in the human population. In the case of human decision making or behavior, the value above which the mountain peak appears tells us what the most common human behavior is.* The low edges (foothills) of the mountain show that fringe behavior is uncommon. What is important here is the width of the mountain. For example, a community of religious zealots will have a narrow mountain, as how to behave has already been decided for you. The narrower the mountain, the more limited is the variety of human behavior, making society more predictable. A society of unpredictable people will surely make the width of this normal distribution wider.†

In this context, we should mention the topic of social physics. The Wikipedia entry on this topic begins by stating that social physics, or sociophysics, is a field of science that uses mathematical tools inspired by physics to understand the behavior of human crowds. In modern commercial use, it can also refer to the analysis of social phenomena with Big Data. As physics in essence is in the business of predicting future outcomes, the mere fact that the social field of research is now named social physics points to the fact that society is becoming increasingly predictable. (In fact, such ideas go back many years.‡)

---

*The center of the mountain points to conformity, or what was once referred to as "normal" behavior. I am relieved that psychologists have stopped using the word normal, and in the context of this book, this may be acknowledged as being a positive sign that society is ready for the change advocated in this book.

†As noted previously, for the lovers of science fiction, let me add that in case Earth is ever attacked by sinister aliens, if humans are able to become less predictable it would be much harder for the aliens to fight humans and conquer us. In such a case, fighting predictability will even ensure biological human life, namely life as we humans have defined it so far.

‡The idea of transferring laws of nature from the realm of physics in order to understand or describe other realms is not new. In 1636, the English philosopher Thomas Hobbes met the astronomer Galileo Galilei, famous for his description of motion, and started to outline how society could be described by laws of motion and the mathematical tools of physics. A similar occurrence can be found in the works of eighteenth-century philosopher David Hume, who was influenced by Newton and talked of "mental forces" and the "force of habit," which is reminiscent of our law of mental inertia. Hume even talked of the force of habit as being analogous to gravity.

Another very important question is what societal frameworks may contribute to the flourishing of unpredictability. As it is much harder to build a functioning society for unpredictable people, the societal and political barriers against unpredictable individuals are enormous. I believe that in the short term, only by showing that unpredictability is in fact creativity, which in turn supports economic growth, will we be able to convince governments, societies, schools, and parents to support the idea of unpredictable people, an assertion that will force change in education and in all walks of life. As noted, this can then be achieved by, for example, redefining happiness and reaching a balanced mind with what concerns logic.

In addition to education that will foster creativity and thus unpredictability, society may even support physiological changes to support such a transformation. As noted in the conclusion of Part V, identifying which parts of the brain or which phenomena in the brain are responsible for unpredictability (e.g., noise) may enable us to increase unpredictability, or in other words, creativity.

Only much later do I expect that the motivation to be alive according to the new definition of life presented in this book will give rise to enough pressure by individuals that societies as a whole will have no choice but to adapt to this new way of thinking and behaving.

But even if pills that can fight predictability do exist, the future of human predictability can still go very badly for society. As noted, when the technology to measure our predictability does mature, commercial companies and social media, homeland security institutions and rulers, may not be in a great hurry to reveal their cards, as they will quite certainly hope that this technology will provide them with an economic, security, or political edge. Revealing that high levels of predictability exist may even be deemed as posing a danger to social stability. So society may not even be aware that it and its individuals are predictable, and consequently, no pills will be designed and consumed. To describe this morbid future, we may coin the phrase *hidden predictability*.

In fact, society should continuously ask the question of whether we could already be living in a state of hidden predictability, namely, a social pyramid of predictability, where each layer of society predicts the layer beneath it, as each layer has more harvesting and predicting technology than the layer beneath it. Whoever is at the very top of the pyramid will of course make sure that the pyramid is stable. Indeed, hidden predictability is a powerful tool for stability, as each layer knows exactly how to provide the needs of the layer beneath it, and if there are rare unsatisfied outliers, they can be easily predicted, identified, and isolated to ensure they don't cause social unrest.[6]

To conclude, it is crucial for society to put in place watchdogs that would monitor the state of all aspects of human predictability, especially the technology to measure predictability. Such watchdogs and think tanks, which could perhaps be formed in the framework of universities, would have the enormous responsibility of, on the one hand, protecting society from the dark side of predictability and, on the other hand, advising society on how to utilize the knowledge regarding predictability to design remedies that will enable individuals and society to transcend, truly thrive, prosper, and *live*.

Epilogue

Dear Eve, years before I met you, I became obsessed with the question "Am I really alive?" Many of us ask ourselves this very question in different versions, such as "What is the meaning of my life?" or "In what sense am I alive?" but we are not really sure what we mean by it. It seems our intuition is telling us that the medical definition of *alive*, based on heart and general brain activity, is too rudimentary. At some point we are content that the mere fact that we are breathing is good enough proof that we are alive, but then we recall that the most primitive worm, fish, or lizard breathes as well, and has a heart as well, and a suspicion crawls in that our contentment was ill conceived.

I was obsessed, but then came life with all of its distractions, and I had all but forgotten my obsession. That is when you appeared with your surprising question, "Is the atom alive?" and resurrected my old passion. Although we only met briefly for a few hours, you later became a strong voice within my mind, helping me sift through the maze of scientific facts, as well as philosophical considerations and arguments.

When you asked the question, I argued that we should first understand what the word *alive* means, and you agreed that by *alive* you were referring to human life. It was then clear that our two questions were one and the same, and we need to understand what *human life* means. As the field of consciousness seems to still be diverging and is not expected to converge any time soon, we had to pivot around a completely different focal point to build our understanding.

In this book we distilled these intuitions into a rigorous reflection leading to a new and objective definition for the level of (mental) human life. Indeed, present theories of consciousness similarly conjecture that different biologically healthy brains will in the far future exhibit differing values for objective evaluations of consciousness. Our new definition, expected to enable some near-future answers, is based on the objective observable, human predictability, P.

Although we are speaking here of mental death rather than biological death, we use the word *death* not as some metaphor or a mere philosophical reflection, because we realize that mental death should be regarded as real death, since mental life is what is unique about human life. In doing so, we could come to the conclusion that an actual experiment on humans, measuring their predictability—an experiment that is now quickly unfolding due to the striking rise of AI and Big Data—is poised to become the most profound experiment ever performed on humans.

We have also examined how the brain gives rise to two forces, one for predictability and the other against it, formidable forces that are in constant armed conflict with each other. These forces drive us humans between the two extremes of a deterministic machine and randomness. It seems that most of us are closer to the predictable, deterministic extreme. Being in the middle between the two extremes is the best we can hope for. Namely, human life is maximized if it is *unpredictable but not random*.

It is important to note that this middle ground of $P = 0.5$ is by no means a static point. It is not stable. It is like standing at the very peak of a mountain, and this peak is narrower than our shoe, and gusts of wind are forever trying to topple us into the abyss. As our environment is fluid and as life is a journey of constant change, aiming for this point of maximal human life is a never-ending battle, a Sisyphean struggle for balance, a frantic movement for life.

To end my answer to you, let me emphasize yet again that as we are typically closer to the deterministic extreme, the main conclusion is that human life is all about distancing ourselves from the machine that is built into us, namely the predictability that is within us. In addition to internal deterministic processes in the brain fixed by nature, life is full of external enhancers of the predictable human. We have briefly examined the conditioning of humans, exemplified in war and happiness as two such enhancers. Logic may also drive predictability, because an external agent can identify what the next logical step is, and because it seems to suppress the ability for imagination and creativity, which require a random seed.

Predictability is the opposite of freedom, or specifically, free will. I have noted that some scientists, after decades of experiments (although not having definite proof), are willing to state that they believe that some form of freedom exists and may even be enhanced through our own actions. We may just as well concede to their belief, because if they do happen to be right, and if we do not act as if we have some ability to enhance our freedom, with all the implications freedom brings, our losses will be immeasurable. Although unpredictability is still a far cry from freedom, I interpret such actions toward increasing our freedom as actions that distance us from the machine, and this may be done either by weakening our deterministic vetoing filters, as described in the two-layer model of the brain, or through increasing randomness by enhancing our neural network noise or quantum dominance.

Noise and randomness may one day be enhanced artificially through medical intervention, including perhaps pharmaceuticals or some other stimuli such as electric shocks or maybe even brain implants. This will be a pill of the second kind. It will be a commercial product that will be made available only once the masses demand it; this will probably happen only after the measurement of human predictability is made readily available for each individual and after humans understand the grave philosophical consequences of being predictable. If we are lucky, this may happen sooner once people understand that fighting predictability means enhancing creativity, and this gives them an economic edge. If we are unlucky, the powers that be will keep hidden the fact that for the most part humans are predictable, and we will never see the need to fight it.

With what concerns the pill of the first kind, namely the kind that people can already design and administer to themselves now, I have discussed ways to fight predictability, such as achieving a balanced mind with regard to logic. We must acknowledge that the logical mind (or exploiting feedback loop) and its high predictability are extremely necessary for our success, both on the unconscious and conscious levels. At the unconscious level, our body retracts our hand when we touch a hot object. This is completely predictable and necessary. At the conscious level, our predictable logical feedback loop enables us to survive automobile driving, or in the extreme case of an aerobatics team, to fly at high speed in close formation, where the tip of your wing is not more than a meter or two from the wing tip of another airplane. And, of course, the logical mind serves us in endless other ways, from planning a budget and designing new medicine to getting to Mars and beyond.

So the logical, predictable mind should not be marginalized; it should be balanced. I have also discussed happiness as crucial in the fight against predictability. As noted, in its simple, common form, it is an enhancer of predictability, nothing less than an agent of the machine. However, on the other hand, by redefining happiness, it may become a powerful tool against predictability, a major ingredient in the pill.

Therefore, at present, in a scenario in which enhancement of freedom, or at the very least, unpredictability, is possible, we must fight for our freedom by identifying predictability enhancers in our life, redefining happiness, and curbing the logical mind by empowering imagination and creativity to go beyond what the logical mind considers logical, to take it outside of its comfort zone, for example, via abstract content (e.g., poetry, dancing, painting), at the very edge of what our logical mind may digest, namely, at the very edge of what it may interconnect with and interpret. As creativity may be introduced into a redefinition of happiness, and as creativity's exploration balances the exploitation of the logical mind, and generally speaking, as creativity is synonymous with unpredictability—the latter being a measure of how far we have distanced ourselves from the machine, creativity is showing itself to be a crucial ingredient of the pill of the first kind, as well as a test of the effectiveness of the pill of the second kind. It, as well as its counterparts such as divergent thinking, are thus a necessary condition for human life no less than oxygen. This must become our conviction, not just in science, technology and the arts, but in every aspect of the human sphere.

I am relieved that psychologists have stopped using the word *normal*, and in the context of this discussion, this may be acknowledged as a positive indication that society is ready for the change advocated in this book. We can only hope that this is a sign that people are ready for the battle to come, a battle which, according to our definition of life, is nothing short of being the most important battle we will ever fight.

Finally, one day, after my answer to you was done, as I was strolling through campus, I thought that you would probably like to have a name for this new living human. So, on a lighter note, I started to wonder: If indeed some freedom, or at the very least, unpredictability, may be acquired, it will clearly bring with it live heroes, the future humans who manage to understand death in this new way and transcend beyond predictability, namely go from P close to one to P close to one-half.* For the love of life, what shall we call such a hero?

A suitable name is perhaps "mindful human," but it has so many connotations going back all the way to Buddha, and then to stress-relief programs in the 1970s, and recently to meditation and well-being.¹ As a name for this human who is transcending, one may also suggest, in humor, "rocket man," as he is flying into the future of our species, but I guess Elton John has taken ownership of this name. Perhaps the best is to use the word *aware*, namely, the "aware person."

So many people have thought about this aware human, for hundreds and thousands of years. Friedrich Schiller, for example, the German poet, talked about the beautiful soul. There was also Hegel who talked about personal development (*Bildung*),² and who said that the challenge of personal growth often involves an agonizing alienation from one's "natural consciousness." Nietzsche, who came a little later, talked about the beyond-man, the *Ubermensch*. (Poor Nietzsche, he had his ideas dragged in the mud into infamy by the Nazis, who thought the beyond-man was a physical being rather than a spiritual one. They were so stupid.†)

But the definitions were so vague. What is "beautiful" and what is "soul"? The aware person presented in this book will be aware of the two extreme edges of randomness and predictability, and of the fact that they serve as

---

*I would venture a guess that in the future they will find a significant correlation between the value of P for an individual and his or her awareness regarding this new kind of death that comes while biologically still existing. If a person embraces the idea that he or she is partially dead although biologically alive, I dare to guess that their predictability will be relatively low, because identifying the enemy is half the battle.

†Indeed, the Nazis took stupidity and the ensuing barbarism to the extreme, and nothing can excuse the pure evil that they have mastered and unleashed, but it is nevertheless insightful to recall that the "science" of racism did not start with them. Following a completely misguided extension of Darwinism to the sphere of human civilization, the eugenics movement appeared. In 1907, the Laboratory for National Eugenics was established at University College London, and in 1912, the first international congress of the eugenics movement was held in London. An interesting anecdote is that Major Leonard Darwin, the son of Charles Darwin, was presiding over the conference, which was dedicated to Sir Francis Galton, a half-cousin of Charles Darwin. Galton, who coined the term *eugenics* in 1883, wrote in 1853 the book on his expedition to Namibia, which, while not directly responsible, can be thought of as providing the ideological roots for the Namibia genocide of 1904 (see footnote in the preface). In 1913 the British Parliament signed onto the Mental Deficiency Act, creating provisions for the institutional treatment of people deemed to be "feebleminded" or "moral defectives." The Americans were not far behind, and in 1910 the Eugenics Record Office was established. By 1935, about thirty-two American states had enacted sterilization laws, and it is believed that some sixty thousand people were sterilized by force. The Belgians and others followed suit. In Germany, in 1933, the Law for the Prevention of Hereditarily Diseased Offspring passed. It is believed that up to 1939, about four hundred thousand people were sterilized by force, and tens of thousands were murdered. As is well documented, this number then rose into the millions by 1945.

archenemies to human life. He or she will thus make every effort to be in the middle, and the success of their effort may be measured and quantified. Obviously, vague terms like *beauty* still have an important role to play, but my claim is that the effort to reach beauty is meaningful only if first and foremost the person is alive according to our new measurable observable. For what is the point of discussing beauty if it is predictable?

—w—

With your permission, I would like to add here one more thought that came to my mind. The British government executed people at the beginning of the nineteenth century for breaking machines. Machine breaking was criminalized already at the beginning of the eighteenth century and then again at the beginning of the nineteenth century. I believe that the people breaking machines mainly feared for their jobs. I hope that after reading my answer to you, you will agree that it is now high time for machine breaking inside of us, for fear of our lives. Luckily, we have a measurable observable to tell us how successful we are in this endeavor.

—w—

Last but definitely not least, I implore you to remain optimistic. The virtual you, roaming around inside my skull, gave me the feeling that you were not.

Throughout human history, there have been many displays of desperation regarding our species. Examples may be found in the Dada movement that emerged in reaction to the First World War, in Voltaire who said that God is a comedian performing in front of an audience too scared to laugh at his jokes, and in Franz Kafka who believed that humans are a fallacy. (I seem to remember that it was Albert Camus who said that Kafka had failed in his mission, as one can still feel between his words a glimmer of hope.*)

In contrast to the words of Ecclesiastes that all is vanity, meaningless, futile, grasping for the wind, I believe that if we distance ourselves from the machine, there is hope for a great transcendence.

Precious Eve, I send you my love. Live long and prosper.

---

*The fact that so many people are still reading Kafka, and the fact that he became famous to the point that in Prague one can find a cafe called Kafka Hummus, should also provide us with some solace. We should meet there one day. . . . Perhaps by then we will already have our P scores.

Appendix A
The Brain

In the following I give a very brief summary of the structure of the brain. You may receive more information from the numerous popular and scientific articles and books written on the brain.[1] I again emphasize that there is no claim being made in this book that I (or anyone else, for that matter) understand how the brain works.[2] But fascinating discoveries are undoubtedly being made (see Appendix B).

The neuron: The neuron is composed of the main body (soma), having a size of some ten to twenty micrometers and holding mainly a nucleus. Out of the soma come numerous branches. The short branches, called dendrites, are used to receive signals from other neurons. A single long branch, called the axon, transfers signals to other neurons and is the essential output of the neuron. It may be hundreds or thousands of times longer than the size of the soma. It ends with the axon terminal, which may have numerous branches connecting to several other neurons. The connections between the axon terminal and the dendrites of the next neuron are called the synapses, and the gap itself is called the synaptic cleft. Neurotransmitters (see following) transmit the signal across this gap. Neurotransmitters are released once an electrical signal called the action potential (see following) traverses through the axon.

Neurotransmitters: The neurotransmitters are made close to the nucleus in the soma and are transported to the synapses through microtubules. Once they cross the synaptic cleft, they bind to a receptor of the postsynaptic neuron. A neurotransmitter influences a postsynaptic neuron in one of three ways:

excitatory (initiates firing or increases firing rate), inhibitory (ceases firing or decreases firing rate), or modulatory (causing long-lasting effects not directly related to firing rate). The two most common neurotransmitters (more than a hundred have been identified) are glutamate and gamma-aminobutyric acid (GABA). Glutamate acts on several types of receptors and has effects that are excitatory at ionotropic receptors and modulatory at metabotropic receptors. Similarly, GABA acts on several types of receptors, but all of them have inhibitory effects. Because of this consistency, it is common for neuroscientists to refer to cells that release glutamate as "excitatory neurons" and cells that release GABA as "inhibitory neurons." Neuromodulators are a bit different, as they are not restricted to the synaptic cleft between two neurons and so can affect large numbers of neurons at once. Neuromodulators therefore regulate populations of neurons, while also operating over a slower time course than excitatory and inhibitory transmitters. Famous neuromodulators are dopamine, which is associated with reward, reinforcement, and motivation, and serotonin, responsible for sleep. It is perhaps worth noting that it is here, in the action of the neurotransmitters, that nerve gas creates its deadly effect.

The action potential: This potential, which is sometimes referred to as the nerve impulse, travels down the axon when the neuron fires. It can travel in only one direction, toward the axon terminal. Rather than being a standard electrical current composed of electron transport, here what propagates is the polarity across the membrane of the axon. We may say that the action potential is reproduced along the axon. Sodium and potassium ions (Na+ and K+) move through gated ion channels, which open and close as the membrane reaches its threshold potential. Whether the impulse is created or not is all about voltage. The baseline is typically at −70 mV (millivolts). If the potential reaches −50 mV, the ions start moving across the membrane, and the potential goes to +40 mV in one millisecond (ms). After another millisecond, the potential goes down to −90 mV. After 2 ms more, it returns to its baseline. Concentration gradients are key behind how action potentials work. In terms of action potentials, a concentration gradient is the difference in ion concentrations between the inside of the neuron and the outside of the neuron (called extracellular fluid). Neurons have a negative concentration gradient most of the time, meaning there are more positively charged ions outside than inside the cell. This regular state of a negative concentration gradient is what causes the negative baseline potential.

Parts of the brain: The brain, made of some one hundred billion neurons and weighing about 1.5 kg, has three main parts: the cerebrum, cerebellum, and brainstem. The cerebrum is the largest part of the brain and is composed of the right and left hemispheres. We have already noted—which may turn out, as with many other ideas about the brain, to be an oversimplification—that the left part of the brain is more tuned to analytical (logical) analysis, as well as controlling speech and writing, while the right part is more intuitive, seeking an understanding through insight. It also controls spatial ability and artistic and musical skills. The two hemispheres are joined by a bundle of fibers called the corpus callosum, which transmits messages from one side to the other. Each hemisphere communicates with the opposite side of the body (smell seems to be an exception). If a stroke occurs on the right side of the brain, your left arm or leg may be weak or paralyzed. The cerebellum is located under the cerebrum. Its function is to coordinate muscle movements and to maintain posture and balance. The brainstem acts as a relay center connecting the cerebrum and cerebellum to the spinal cord. It performs many automatic functions such as breathing, heart rate, body temperature, wake and sleep cycles, digestion, sneezing, coughing, vomiting, and swallowing.

In an effort to simplify the understanding of the brain, many specialized areas have been identified. For example, the cortex is the outermost layer of brain cells. Thinking and voluntary movements begin in the cortex. The external part of the cortex holds the neurons' cell bodies and dendrites and is called gray matter, while the internal part holds the axons, called white matter. Another rather popular division describes four lobes: the frontal lobe controls thinking, planning, and organization, as well as short-term memory and movement; the parietal lobe processes what you feel in terms of taste, texture, and temperature; the occipital lobe processes images from your eyes and compares them to stored images in order to achieve visual recognition; and the temporal lobe manages your emotions, processes sensory information (e.g., from your ears), stores and retrieves memories, and understands language.

Finally, just to show how complex things really are, let me note that beyond the above description, there are many more types of cells required for the operation of the brain. For example, no one still really knows the full expanse of the functionality of the Astrocytes. These are star-like cells that are probably just as abundant as neurons and their exact contribution is still under investigation.[3]

—∞—

For more information beyond this extremely brief presentation of the different parts of the brain, especially on the production of will, I refer the reader to the book *The Will and Its Brain* to show just how complex the manufacturing of will is in the human brain.[4]

Appendix B
Attempts to Quantify Consciousness and Self-Awareness

In this appendix, we dive deeper into the attempts that have been made to quantify and measure consciousness and self-awareness. As the definitions of these concepts vary considerably—hence my description of them as being vague—and as I believe this will continue to be the case in the foreseeable future, better measuring machines probing the brain will not, in my view, help resolve the hard problem of consciousness.* Nevertheless, it is informative to briefly review the heroic efforts that have been made by the different disciplines of science.

Let me start by quoting from *The International Dictionary of Psychology* a statement by Stuart Sutherland (1989): "Consciousness is a fascinating but elusive phenomenon; it is impossible to specify what it is, what it does, or why it evolved. Nothing worth reading has been written on it." I must say that, personally, I feel these words are much too harsh, but still it is clear that con-

---

*The easy problems of consciousness are those that seem directly susceptible to the standard methods of cognitive science, whereby a phenomenon is explained in terms of computational or neural mechanisms, for instance, explaining how the brain integrates information, categorizes and discriminates environmental stimuli, or focuses attention. The hard problem of consciousness (David Chalmers, "Facing Up to the Problem of Consciousness," *Journal of Consciousness Studies* 2 [1995]: 200–19) is the problem of explaining the relationship between physical phenomena, such as brain processes, and experience (i.e., phenomenal consciousness, or mental states/events with phenomenal qualities or qualia). Why does a given physical process generate the specific experience it does—why an experience of red rather than green, for example? http://www.scholarpedia.org/article/Hard_problem_of_consciousness. See also a recent review: Robert L. Kuhn, "A Landscape of Consciousness: Toward a Taxonomy of Explanations and Implications," *Progress in Biophysics and Molecular Biology* 190 (2024): 28-169.

sciousness is a very hard problem, which will not be solved any time soon.[1] In any case, such words have also clearly driven this field into an experimental phase, and for that they should be commended. When did the journey to understand consciousness start? It's a matter of your point of view, but one may, for example, say that the empirical journey began in earnest following the seminal paper by Francis Crick and Christof Koch in 1990.[2] An important step was made in the identification of the neural correlates of consciousness (NCC), which constitute the minimal set of neuronal events and mechanisms sufficient for a specific conscious perception.[3] In the last thirty years, there has been an explosion of new theories and empirical studies.[4]*

In my view, the first hard problem in identifying and quantifying consciousness is that it inhabits the same physical space and the same elementary physical processes (e.g., neurons, neurotransmitters) as unconsciousness. So the difference is clearly about the processing, which is more subtle to probe and understand. Freud used the iceberg analogy to explain that consciousness is just a small part (tip of the iceberg) of the mental operations that exist inside the skull. Perhaps one of the biggest clues at our disposal is that a big chunk of the brain, the cerebellum, did not develop any consciousness.† Another clue, discussed earlier, is the chemicals with anesthetic capabilities that can turn consciousness off. The second hard problem, in my humble opinion, has to do with the fact that we are trying to examine here an objective reality through subjective means. The objective reality is our brain and its processes, but consciousness exists so far only in the subjective experience of each individual and may be different for every person. A very rudimentary but insightful example of the idiosyncrasy of our subjective mental state may be found in the dress photograph that became a viral internet phenomenon in 2015. The dress was thought to have different colors by different people, which is indicative of different processing. Perhaps future brain probes will be able to objectively quantify consciousness, but this is very far away.

---

*Nevertheless, some scientists continue to be extremely frustrated with the state of the field, to the point that they feel a completely new approach is needed, one that is even much more open to things still considered metaphysics. See, for example, Princeton Engineering Anomalies Research (PEAR) and International Consciousness Research Laboratories (ICRL): http://icrl.org; https://opensciences.org/organizations/international-consciousness-research-laboratories.

†Trying to define the border between what is a conscious act and what is not is not easy. See, for example, Daniela R. Waldhorn, "What Do Unconscious Processes in Humans Tell Us about Sentience?" (2019), https://rethinkpriorities.org/publications/what-do-unconscious-processes-in-humans-tell-us-about-sentience.

To appreciate the huge variance in the definitions of the different terms currently used to describe human life, one may start with the phrase *sentient beings*, whereby even lobsters and crabs were recently added to the list of the British Parliament applicable to an animal welfare bill, based on scientific findings by the London School of Economics. Hence, using this phrase to try to define human life is rather lacking in specificity. A review paper was published in 2013 in *Animals* by Helen S. Proctor and colleagues titled "Searching for Animal Sentience: A Systematic Review of the Scientific Literature."

The term *self-awareness* is just as complex and lacking in differentiation powers for the unique case of humans. For example, the mirror test is typically used to measure self-awareness, including the mark test, which involves animals spontaneously touching a mark on their body that would be difficult to see without the mirror. Chimpanzees, elephants, dolphins, and even birds have been shown to have self-awareness. Self-awareness may be divided into bodily, social, and introspective, where the latter seems to be the most advanced.[5] Some claim (e.g., David DeGrazia [2009]) that there is independent evidence from metacognition studies involving monkeys that certain nonlinguistic creatures are also introspectively aware.

In our brief scan of the terms used to define human life thus far, we now come to the holy grail, namely, consciousness. Philosophers and scientists have been debating its operational meaning for centuries. The Wikipedia entry is rather informative, and I will not copy/paste it here. Let me just quote two sentences stating that "experimental research on consciousness presents special difficulties, due to the lack of a universally accepted operational definition," and, "For many decades, consciousness as a research topic was avoided by the majority of mainstream scientists, because of a general feeling that a phenomenon defined in subjective terms could not properly be studied using objective experimental methods." Nevertheless, it would be misleading not to mention the heroic advances being made by researchers. See, for example, a special issue of the journal *Neuron*[6] and the optimistic view of some neuroscientists.[7]

Maps of the wide variety of approaches to consciousness were published, for example, in *Frontiers of Psychology* and *Progress in Biophysics and Molecular Biology*.[8] The former work starts by defining five fundamental categories in which questions about consciousness are classified: definitional (definitions of key concepts relating to our subject matter), phenomenological (examining from a subjective point of view of personal experiences), epistemological (theory of knowledge and the distinction between justified belief and opinion), ontological (a branch of philosophy that studies concepts such as existence, being, becoming, and reality), and axiological (the philosophical study of values, such as right and wrong, good and bad). This work attempts to compare integrated information theory[9] and global workspace theory.[10]

Dehaene and colleagues[11] proposed a neuronal version of the latter theory, the global neuronal workspace (GNW), which has become one of the major contemporary neuroscientific theories of consciousness. According to GNW theory, conscious access occurs when incoming information is made globally available to multiple brain systems through a network of neurons with long-range axons densely distributed in prefrontal, parieto-temporal, and cingulate cortices.

We should pay special attention to the phenomenological point of view. Whereas this is a simplifying point of view in physics, where the observation of phenomena with instruments creates a true public language in the sense that for everyone the spoken name points to the same content, in the examination of consciousness, each mind has its own subjective phenomenological point of view. There is no true public language. This is why philosophers were so interested in the sensation of pain. Only one mind really feels it, and the others feel his or her pain only through the linguistic representation that he or she, who is suffering from the pain, produces. As noted in the main text, this problem was eloquently put forward by Ludwig Wittgenstein and his thought experiment which he called "the beetle in the box." Wittgenstein invites readers to imagine a community in which the individuals each have a box containing a "beetle." No one can look into anyone else's box, and everyone says he knows what a beetle is only by looking at *his* beetle. If the "beetle" had a use in the language of these people, it could not be as the name of something—because it is entirely possible that each person had something completely different in their box, or even that the thing in the box constantly changed, or that each box was in fact empty.

Concerning integrated information theory (IIT), two recent papers[12], one popular and one more scientific, discuss the epistemological foundation. Both seem to reaffirm the notion that while the quality and quantity of data coming from machines like fMRI and EEG are improving, there is still plenty of, at times acrimonious, confusion and debate.

The first paper focuses on IIT, invented by Giulio Tononi.* This theory has a mathematical value for consciousness represented in the theory by the Greek letter *phi*. That is, the hope is to be able to evaluate the level of consciousness with a number. Inspired by IIT, an empirical test for consciousness was developed by Marcello Massimini (in collaboration with Tononi). At present, Massimini uses transcranial magnetic stimulation (TMS) of the brain (instead of pictures or verbal commands) and measures the activity following the stimulation with electroencephalography (EEG), creating a value called the "perturbational complexity index" (PCI). According to Tononi, PCI is a poor man's phi. Namely, it is still not the ultimate quantification of consciousness through the measurement of phi, but it is on the way. This paper quotes Scott Aaronson as stating that the theory is wrong. For a recent test of the theory, see the work of Karim Jerbi and Jordan O'Byrne.[13]

The second paper describes empirical data coming from work on people with brain injuries, or with psychiatric disorders, and even people in a coma. Such work, intended primarily to help repair impaired brains and to communicate with unresponsive patients, might also offer a window to the neural signatures of consciousness. The article ends by quoting Matthias Michel as stating that IIT is wrong. Indeed, in 2023, 151 scientists from 150 institutions signed a letter stating that in their opinion IIT is still to be considered pseudoscience.[14]

A comparison between GNW and IIT is given by Melloni et al.[15] This article states that the differentiating tests now "focus on two key questions: Where are the anatomical footprints of consciousness in the brain: Are they located in a posterior cortical 'hot zone' advocated by the IIT, or is the prefrontal cortex necessary as predicted by the GNW (or GNWT)? And, how are conscious percepts maintained over time: Is the underlying neural state maintained as long as the conscious experience lasts, in line with IIT, or is

---

*IIT attempts to identify the essential properties of consciousness (axioms) and, from there, infers the properties of physical systems that can account for it (postulates). Based on the postulates, it permits us in principle to derive, for any particular system of elements in a state, whether it has consciousness, how much, and which particular experience it is having. The postulates are represented mathematically, which allows us to calculate the consciousness value of a system: phi. Some IIT principles about conscious experience: it's structured (apple on a table has color, position, texture), it's informational (green apple is different from red apple or lack of apple), it's integrated in one whole experience (you don't perceive the apple separately from its color or from the loud music from the neighboring apartment). IIT principles and the consciousness value phi aim to represent both subjective experience (= hard problem, phenomenology) and empirical evidence (= easy problem). http://www.scholarpedia.org/article/Integrated_information_theory.

the system initially ignited and then decays and remains silent until a new ignition marks the onset of a new percept, as GNWT holds?" (Concerning the concept of ignition, which we also encountered in the section on noise, see also Stanislas Dehaene and Jean-Pierre Changeux's work in *Neuron*.[16])

In my humble opinion, as we still know so little, it is highly unlikely that one of these theories is complete. Still, having adversary theories is extremely helpful in designing differentiating tests, which will then hint at how to make further progress in the right direction. I should also note that there are many more interesting opinions that oppose the general way of thinking in this field, opinions that should also be considered.[17]

Finally, see also a unique plea by numerous neuroscientists concerning the future of the field.[18]

Let me end this brief review of the field of consciousness by noting that the field still seems to be plagued by diverging opinions and hard problems, so the fundamental premise of this book is warranted, namely that, if we want any near-future answers, even if partial, we need to also utilize parallel tracks (in theory and experiment) not based on the term *consciousness*.

Appendix C
Research Dealing with Prediction of Human Behavior

In the process of collecting information about humans, and consequently analyzing the information and predicting the human, there are two major players, the collection and the analyzing. Collecting the information will of course become increasingly easy and commonplace. We are in a platform economy. Even an agricultural company like John Deere has put algorithms that harvest information on its tractors. Obviously, there will be increasing efforts by regulators (e.g., General Data Protection Regulation [GDPR]) to contain this digital pandemic, but in the race between technology and regulators, I think we all know who will eventually win.

The analyzing part is also growing exponentially in power. In fact, one may view the enormous effort by experts to regulate the development of AI as a sign of how quickly the experts believe the technology will develop (if unregulated). For example, one may follow the very difficult negotiations surrounding the EU AI Act.[1] On December 8, 2023, one milestone agreement, among many that will be required in the future, was reached.[2] But again, as with the collection of data, so with analyzing it by AI—I believe we all know that the regulators will eventually lose.

Today's social media companies are afraid of the public wrath concerning their harvesting of information for their recommendation systems, to the point that the once forbidden word *f**k* has now been replaced in the corridors of algorithm writers by the forbidden word *feelings*; you are not allowed to talk of knowing and predicting the feelings of an individual. But what is

227

now socially and even politically correct will not stop the platform economy. Specifically, it stands to reason that individuals will have no choice but to agree to give up their right to privacy in order to have AI and deep-learning machines help them in managing more effectively their well-being, including their career, personal life, health, security, and even simple home comfort (so-called AI personal assistants). I am quite sure that in our home, in the not so far future, not only electronic and electrical appliances but even the furniture will harvest information (e.g., the smart home concept). As noted in the main text, data harvesting in the near future will also include wearable devices. In fact, data harvesting will even include direct access to the brain (BCI technology, i.e., brain-computer interface), whether via implants or external probes (see, for example, the wristband that was built by a startup named CTRL-labs, a company that was acquired by Meta in 2019). The expected capability of BCIs to harvest information was also discussed in the book *The Battle for Your Brain* by Nita Farahany. I am therefore confident that the harvesting of information covering all aspects of our personality will grow exponentially.

The main challenge of being able to predict a decision taken by a human at a specific moment is the correct understanding by AI of the context in which the human was engulfed at the time of the decision making. Context typically involves physical aspects, relationships, and intent. All three topics are hard. The physical aspects include, for example, the state of your body and present actions, as well as your physical location in space. The former will be addressed by wearable sensors, while these will also be able to address the latter through real-time mapping applications. For example, a development effort focuses on simultaneous localization and mapping (SLAM). Relationships include relationships with objects around you as well as with animals and people. Here, first-person data is extremely valuable, and a growing number of R & D programs are focused on that (e.g., Ego4D and the Aria project [both from Meta]). Intent is probably the hardest of the three. But taking into account the vast knowledge that AI will have concerning who you are, this challenge seems to be indeed surmountable. For early work on context, see the article by Wanyi Zhang and colleagues in *EPJ Data Science*.[3] The emerging field of erobotics, in which AI is geared for developing emotional and sexual bonds with humans, will clearly push the understanding of all aspects of context to currently unimaginable capabilities.

The important question remains: Given a future incorporating an almost infinite volume of collected information (the Big Data of today is just the seed of a giant tree) and almost unlimited memory and computing capacity, and given better and better models of personality types and traits, as well as collaborative filtering, how good will the predictive algorithms be? If we are indeed to a large extent deterministic, will these algorithms be able to prove with a surprising level of success that we are predictable?

In the Wikipedia entry on predictability, one may find the following sentence: "Another example of human-computer interaction are computer simulations meant to predict human behavior based on algorithms. For example, MIT has recently developed an incredibly accurate algorithm to predict the behavior of humans. When tested against television shows, the algorithm was able to predict with great accuracy the subsequent actions of characters. Algorithms and computer simulations like these show great promise for the future of artificial intelligence."

AI is of course making its very first steps, and its quality will grow tremendously, but it is still useful to look at where we stand now. Trying to map its current rudimentary capabilities is not an easy task, as many companies are not motivated to reveal their advances, since they would like to keep an economic and even political edge, and also for fear of public uproar.

For completeness, let me now recall some of the examples given in the main text of the book and, where available, introduce additional information. Let me begin by stating that although some companies are advertising predictability scores above 90 percent, this is still a far cry from a general proof of predictability. Proof of predictability would require an academic study that shows predictability on the full range of human activities. However, the preliminary results already obtained exhibit a significant indication.

Companies like Behavio, which came out of the MIT Media Lab and was absorbed by Google in 2013, deal, according to some sources, with "profiling and predicting human behavior." Another specific example is a company named Affective Markets, which follows the same line of thought as US Patent 7,075,000 claiming a system and method for prediction of musical preferences. In fact, still unpublished is their surprising claim for cross-discipline predictability, namely that through musical preferences they can score high on predictability of other human traits. They are already claiming a predictability score of 80 percent. Obviously all these methods rely heavily on deep-learning machines as well as mathematical equations.

Predictability is also being tested by people like John Gottman, this time regarding divorces. Such studies already claim a predictability rate of 80–95 percent (in a 1992 paper with colleagues, they report a 94 percent success rate in predicting divorce with a discriminant function[4]). For example, in an article that may be found on the Gottman Institute website, they claim to use "nonlinear difference and differential equations," and furthermore that "the equations represent a new language for analyzing and understanding couple interactions," and finally that the "theory is designed to be totally disconfirmable, subject to empirical testing."

A company called Pure Matching describes their six-dimensional patented model based on the famous HEXACO personality model; it notes work by scientists from MIT, VU Amsterdam, and the University of Calgary; and it claims that "no AI system in the world can do such a cross-match."

A company called TestColor claims to have an AI engine with a prediction rate above 94 percent. Another company based on visual tests is VisualDNA. They claim to have Big Data experts alongside scientists from University College London, Cambridge University, and Columbia University. While I could not find numbers for prediction success rates, they certainly claim strong predictive powers (see their paper on using visual questionnaires to measure personality traits).

Another example is Mind Strong, which claims to use powerful machine learning methods to show that specific digital features (which they call "digital biomarkers") correlate with cognitive functioning, clinical symptoms, and measures of brain activity and allow prediction of mental performance. They quote, for example, a 2018 paper titled "Digital Biomarkers of Cognitive Function."

A new interdisciplinary field called digital phenotyping, utilizing available data from personal digital devices to profile and predict people, is expected to greatly enhance predictive reliability. Companies such as Explorium will optimize the search for data in order to improve the predictability of humans. It goes without saying that dark agents are already doing their best to collect as much data as possible and predict future behavior, for all kinds of malicious reasons. However, in this appendix we are not interested in analyzing intent but rather capability.

Social media companies (e.g., Facebook [now Meta], Instagram, Twitter [now X], TikTok, Netflix, Spotify, YouTube, and even news organizations) are becoming extremely good at predicting what content you will engage with, and for the most part, the algorithms for deciding which posts to show us (which they call recommender systems) have been optimized for that (for a general overview, see, for example, articles by Jonathan Stray). These algorithms can see exactly what content you spend time on or react to (with likes, comments, resharing, or just clicking through and dwelling on content for some time). Even you, the user, cannot describe your behavior as well as they can. This data is fed into deep-learning algorithms that use a huge amount of information with billions of data points to make predictions and output the next items. I have not found numbers stating how good they are. Perhaps they themselves are afraid to publish such results. The industry is even considering putting all the data collected by the different companies in one place. Of course, the idea is that this very secure place will be owned by the user, and it will only be used for the user's benefit, but we are not here to discuss the heroic efforts of good-willed, conscientious AI developers to keep social media ethical (an effort they call "value alignment"[5]) but rather to examine how far they have gone and how far they will go in the future in proving our conjecture of predictability.

Present-day academic articles and theses having titles such as "Predicting Human Preferences from User-Generated Content," "A Machine-Learning Approach for Predicting Human Preference . . . ," "Predicting Human Preferences Using . . . ," "Choosing Prediction over Explanation in Psychology: Lessons from Machine Learning," and "A Big Data Revisit to Fundamentals of Personality Psychology"[6] (these are real titles of existing papers and theses) will clearly become more and more commonplace as industry and the state understand the potential and start investing heavily in related research.

As noted in the main text, a 2010 Northeastern University popular statement declared that "human behavior is 93 percent predictable, a group of leading Northeastern University network scientists recently found." This is of course very far from predicting humans, but still, it is indicative of the trend.[7]

Will we get to 100 percent predictability on all fronts of human activity? My conjecture is that for some people we will get very close, and I hope that when the time comes, we will indeed be told of this achievement. Commercial interests and national security arguments, and even the wish for political control, may strongly oppose making this public.

Appendix D
Kitsch and the Human Satisfaction Resonance

So, are standard aesthetics, which may also be called idle aesthetics, and kitsch somehow similar? Yes, I think that indeed they are, in the sense that the reaction is predictable. In the battle against predictability, we need to fight both.

It's all about a resonance condition.

Every object has its mechanical resonance frequency. It's the frequency in which it can absorb the most energy, and for which it vibrates the strongest, like the unique frequency each string on a guitar or piano or violin has. For example, it's very dangerous for a bridge if a group of soldiers marches on it at its resonance frequency, as the bridge may absorb enough energy to collapse.

It seems the human soul also has its resonances, and standard aesthetics and kitsch are simply engineered to work on the most common frequencies so that they strongly interact with our deepest feelings.

I think it was Milan Kundera who said that kitsch is the second tear drop. I once had a meeting in Prague, in a cubist café, with a guy who teaches aesthetics. His name is Tomas Kulka. He said that for something to be kitsch, it needs to satisfy three conditions. First, the described situation has an emotional appeal; namely, it is known to arouse emotions. Second, the subject and technique never wander into unchartered territory; that is, the situation will seem familiar to the observer or reader and create intimacy. And third, the exact details are irrelevant and may be changed without hindering the effect.

Indeed, in the context of the chemical machine, aesthetics and kitsch seem quite similar, as both are essentially fraudulent. A person looks at something, but he really does not see it, in the sense that seeing is observing and observing is analyzing in a complex way for which the final verdict is unknown. With both of them, if they are good representations of what they purport to be, the verdict is known in advance, and very quickly the person receives a deep sense of pleasure.

Standard aesthetics and kitsch seem to be a trivial action upon the machine in the human, aiming at the lowest-order (simplest, most common) satisfaction resonance. In other words, the lowest common denominator resonance in humans.

Notes

PREFACE

1. Stuart Sutherland, *The International Dictionary of Psychology* (1989).

2. https://en.wikipedia.org/wiki/Midjourney.

3. https://www.discovermagazine.com/technology/4-robots-that-look-like-humans.

4. https://www.nytimes.com/2023/09/20/books/authors-openai-lawsuit-chatgpt-copyright.html.

5. https://futureoflife.org/open-letter/pause-giant-ai-experiments.

6. https://www.microsoft.com/en-us/research/publication/sparks-of-artificial-general-intelligence-early-experiments-with-gpt-4.

INTRODUCTION

1. Daniel Kahneman, Olivier Sibony, and Cass R. Sunstein, *Noise: A Flaw in Human Judgement* (2021).

2. https://www.theguardian.com/books/2021/aug/25/being-you-by-professor-anil-seth-review-the-exhilarating-new-science-of-consciousness.

3. On the science behind mind-altering drugs, see popular articles such as Charlotte McAdam, "Investigating the Profound and Bizarre Link between Creativity, Psychedelics and Music" (2020); Holly Williams (BBC), "How LSD Influenced Western Culture" (2018); Renee Deveney, "5 World-Famous Artists and Their

Drugs of Choice" (2021); John Sewell, "7 Famous Artists Who Experimented with Narcotics" (2020); Dirk Hanson, "Eye on Fiction: Heavenly and Hellish—Writers on Hallucinogens" (2014). See a 1989 experiment with painters: Oscar Janiger and Marlene Dobkin de Rios, "LSD and Creativity," *Journal of Psychoactive Drugs* 21 (1989): 129–34. See also more detailed books like *The Eureka Factor* by Mark Beeman and *How to Change Your Mind* by Michael Pollan.

4. R. Gordon Wasson, Albert Hofmann, and Carl A. P. Ruck, *The Road to Eleusis* (1978); Brian C. Muraresku, *The Immortality Key* (2020).

5. Marcus Du Sautoy, *The Creativity Code*, Belknap Press (2019). See also "Photo Vogue Festival 2023: What Makes Us Human? Image in the Age of AI," https://www.vogue.com/video/watch/what-makes-us-human.

6. https://en.wikipedia.org/wiki/Midjourney.

7. Ravinder Jerath et al., "Meditation Experiences, Self, and Boundaries of Consciousness," *International Journal of Complementary and Alternative Medicine* 4 (2016): 00105.

8. Michael M. Schartner, Robin L. Carhart-Harris, Adam B. Barrett, Anil K. Seth, and Suresh D. Muthukumaraswamy, "Increased Spontaneous MEG Signal Diversity for Psychoactive Doses of Ketamine, LSD and Psilocybin," *Scientific Reports* 7 (2017): 46421, https://doi.org/10.1038/srep46421. On ketamine, see also Duan Li and George A. Mashour, "Cortical Dynamics during Psychedelic and Anesthetized States Induced by Ketamine," *NeuroImage* 196 (2019): 32–40. See also Tim Bayne and Olivia Carter, "Dimensions of Consciousness and the Psychedelic State," *Neuroscience of Consciousness* (2018): niy008; Martin Fortier-Davy and Raphael Milliere, "The Multi-dimensional Approach to Drug-Induced States: A Commentary on Bayne and Carter's 'Dimensions of Consciousness and the Psychedelic State,'" *Neuroscience of Consciousness* (2020): niaa004; Raphael Milliere et al., "Psychedelics, Meditation, and Self-Consciousness," *Frontiers in Psychology* (2018). Finally, see also the books: Andrew Weil, *The Natural Mind: A New Way of Looking at Drugs and the Higher Consciousness* (1972), and Mitch Earleywine, *Mind Altering Drugs: The Science of Subjective Experience* (2005).

9. Kahneman et al., *Noise*.

CHAPTER 1: THE NEAR FUTURE WILL CHANGE EVERYTHING

1. Christoph Marty, "Darwin on a Godless Creation: 'It's Like Confessing to a Murder,'" *Scientific American* (2009), https://www.scientificamerican.com/article

/charles-darwin-confessions; see also Janet Browne, *Darwin's "Origin of Species,"* Atlantic Books (2006); Steve Jones, *Darwin's Ghost: The Origin of the Species Updated*, Ballantine Books (2001).

2. See, for example, https://en.wikipedia.org/wiki/Torches_of_Freedom.

3. See, for example, https://aninjusticemag.com/this-is-bananas-a-fruit-company-coup-and-the-rise-of-consumer-culture-20053aab2c40.

4. See, for example, the compelling BBC series *The Century of the Self.*

5. https://www.theverge.com/2013/4/12/4217618/google-purchases-behavio-a-startup-that-makes-predictions-based-on-smartphone-data.

6. To get a glimpse of the future pyramid of predictability, one may read an article in the New York Times about how China is already predicting the future behavior of outliers: "How China Is Policing the Future," *New York Times*, June 25, 2022.

7. For a general view, see, for example, Chapter 2 in *Recommender Systems Handbook*, 2nd edition, ed. Francesco Ricki, Lior Rokach, and Bracha Shapira, Springer (2015).

8. Jonathan Stray, Alon Halevy, et al., "Building Human Values into Recommender Systems: An Interdisciplinary Synthesis," ArXiv (2022), https://arxiv.org/pdf/2207.10192.

9. See, for example, Albert-László Barabási, "You're So Predictable," *Physics World*, 2010. See also M. C. Gonzales et al., "Understanding Individual Human Mobility Patterns," *Nature* 453 (2008): 779–82; E. Cho, S. Myers, and J. Leskovec, "Friendship and Mobility: User Movement in Location-Based Social Networks," in *ACM KDD*, San Diego, CA (2011), https://dl.acm.org/doi/10.1145/2020408.2020579.

10. For the full story regarding the Replika app and erobotics, see the chapter on intimacy in the book (expected to come out soon) by Alon Halevy and colleagues, with the tentative title *The AI Partner.*

11. Peter A. Biro and Bart Adriaenssen, "Predictability as a Personality Trait: Consistent Differences in Intraindividual Behavioral Variation," *American Naturalist* 182 (2013): 621–29. Scientists have also tried to put limits on the achievable predictability in specific scenarios, using a physical parameter quantifying disorder called entropy, or complexity, but I have not found any universally applicable bound; Priit Jarv et al., "Predictability Limits in Session-Based Next Item Recommendation," *Proceedings RecSys* 19 (2019): 146–50, https://doi.org/10.1145/3298689.3346990;

W. Bialek, I. Nemenman, and N. Tishby, "Predictability, Complexity and Learning," *Neural Computation* 13 (2001): 2409–63.

12. There are so many books about the brain that it is hard to choose, but here are a few examples: Frank Amthor, *Neuroscience for Dummies*, 2nd edition (2016); Rita Carter, *The Human Brain Book* (2019); David Eagleman, *The Brain: The Story of You* (2017); Grace Lindsay, *Models of the Mind* (2021). See Appendix A for a brief summary.

13. See, for example, Max Tegmark, "Importance of Quantum Decoherence in Brain Processes," *Physical Review E* 61 (2000): 4194–206. Note that Tegmark calculates the decoherence of a superposition of millions of ions, which indeed is the standard picture of what happens when a neuron fires. He also calculates the decoherence rate of a mechanism suggested by Penrose in which quantum signals of polarization and kink-like excitations propagate through microtubules. He finds very short coherence times.

14. It is safe to assume that the study of quantum theory will, in the future, teach us a lot about our brain and consequently who we really are. To me this is already a good enough reason to invest in quantum computing technologies, even if their practical impact on our life is yet unclear, simply because they will teach us a lot about quantum theory.

I briefly add as a side note that some people even say that the quantum computer may teach us new things about the argument of entropy which I mention in the next page.

On recently funded research on the fundamental connection between quantum theory and the philosophy of the mind see for example: https://fqxi.org/community/articles/display/266?utm_campaign=FQXi%27s%20Reality%20Bites&utm_medium=email&utm_source=Revue%20newsletter.

15. Nicholas C. Stone and Nathan W. C. Leigh, "A Statistical Solution to the Chaotic, Non-hierarchical Three-Body Problem," *Nature* 576 (2019): 406–10.

16. Na Li et al., "Nuclear Spin Attenuates the Anesthetic Potency of Xenon Isotopes in Mice: Implications for the Mechanisms of Anesthesia and Consciousness," *Anesthesiology* 129 (2018): 271–77. For the story of lithium, see Jennifer Ouellette, "A New Spin on the Quantum Brain," *Quanta* magazine, November 2, 2016. For the latest work from the Fisher group on lithium, see Aaron Ettenberg et al., "Differential Effects of Lithium Isotopes in a Ketamine-Induced Hyperactivity Model of Mania," *Pharmacology Biochemistry and Behavior* (2020), https://doi.org/10.1016/j.pbb.2020.172875.

17. For example, see V. Pechuk et al., "Reprogramming the Topology of the Nociceptive Circuit in *C. elegans* Reshapes Sexual Behavior," *Current Biology* 32 (2022): 4372–85, https://wis-wander.weizmann.ac.il/life-sciences/down-synapse-connecting-brain-circuits-behavior; see also Adam Haber and Elad Schneidman, "Learning the Architectural Features That Predict Functional Similarity of Neural Networks," *Physical Review X* 12 (2022): 021051.

18. Karl Friston, "The Free-Energy Principle: A Rough Guide to the Brain?," *Trends in Cognitive Sciences* 13 (2009): 293–301.

19. Dimitri van de Ville et al., "When Makes You Unique: Temporality of the Human Brain Fingerprint," *Science Advances* 7 (2021): abj0751.

CHAPTER 2: COMMON ARGUMENTS AGAINST THE POSSIBILITY OF PREDICTABILITY

1. P.-M. Binder, "The Edge of Reductionism," *Nature* 459 (2009): 332–33; Sophia Kivelson and Steven A. Kivelson, "Defining Emergence in Physics," *NJP Quantum Materials* 1 (2016): 16024; Adam Frank, "Reductionism vs. Emergence: Are You 'Nothing but' Your Atoms?" (2021), https://bigthink.com/thinking/reductionism-vs-emergence-science-philosophy; Paul Humphreys, *Emergence* (2016); *The Routledge Handbook of Emergence*, ed. Sophie Gibb, Robin Findlay Hendry, and Tom Lancaster (2019).

2. M. Gu, C. Weedbrook, A. Perales, and M. A. Nielsen, "More Really Is Different," *Physica D* 238 (2009): 835–39. The fact that in 2021, eleven years since the article was written, only sixty citations have been made could mean that not many scientists or philosophers believe that a significant groundbreaking proof was made here.

3. P. W. Anderson, "More Is Different," *Science* 177 (1972): 393–96.

4. See also seminar by John Krakauer: https://www.youtube.com/watch?v=oo8fKADoqkU.

5. See, for example, "Decision-Making in the Brain," by Dr. Anne Churchland, https://www.ibiology.org/neuroscience/decision-making. Note that they state that there is not much difference in the basic decision-making processes between humans and rats, hence my mention of lizards a few lines above. Note that here they also use a measuring technique I have not mentioned thus far, and this is two-photon imaging, whereby the neuron also emits light when it fires. See also a review in F. Najafi and A. K. Churchland, "Perceptual Decision-Making: A Field in the Midst of a Transformation," *Neuron* 100 (2018): 453–62.

6. Strictly speaking, there can be no "equilibrium" in a dynamic system, so I use this word rather loosely. One can instead use chemical balance or optimized chemical configuration. Chemical balance in the brain motivating actions is more scientifically referred to as the reward system or pleasure centers, or hedonic hotspots. See, for example, K. C. Berridge and M. L. Kringelbach, "Pleasure Systems in the Brain," *Neuron* 86, no. 3 (2015): 646–64; J. M. Richard et al., "Mapping Brain Circuits of Reward and Motivation," *Neuroscience & Biobehavioral Reviews* 37 (2013): 1919–31; Daniel C. Castro and Kent C. Berridge, "Opioid and Orexin Hedonic Hotspots in Rat Orbitofrontal Cortex and Insula," *PNAS* 114 (2017): E9125–34; M. L. Kringelbach and K. C. Berridge, "The Joyful Mind," *Scientific American* 307, no. 2 (2012): 44–45.

7. Eddy Nahmias, "Why We Have Free Will," *Scientific American*, January 1, 2015; Marcel Brass et al., "Why Neuroscience Does Not Disprove Free Will," *Neuroscience & Biobehavioral Reviews* 102 (2019): 251–63; J. T. Ismael, *How Physics Makes Us Free*, Oxford University Press (2016). See also the work of Liad Mudrik's group, e.g., Uri Maoz et al., "Neural Precursors of Decisions That Matter—an ERP Study of Deliberate and Arbitrary Choice," *eLife* 8 (2019): e39787.

8. Giulio Tononi et al., "Only What Exists Can Cause: An Intrinsic View of Free Will," ArXiv (2023), https://arxiv.org/pdf/2206.02069.

9. For a general overview of how mathematics serves as the basis for quantifying brain processes, see Grace Lindsay, *Models of the Mind* (2021).

10. Andrea Eugenio Cavanna, Andrea Nani, Hal Blumenfeld, Steven Laureys, *Neuro Imaging of Consciousness*, Springer (2013).

11. Sean M. Carroll, "Physics and the Immortality of the Soul," *Scientific American*, Guest Blog (2011).

12. Julien Musolino, *The Soul Fallacy: What Science Shows We Gain from Letting Go of Our Soul Beliefs*, Prometheus (2015).

CHAPTER 3: THE HARD ROAD TO A NEW UNDERSTANDING

1. See also David Brooks, "Is Self-Awareness a Mirage?," *New York Times*, 2021.

2. Helge Kragh, "Physics and the Totalitarian Principle," ArXiv (2019), https://arxiv.org/abs/1907.04623; also see https://en.wikipedia.org/wiki/Totalitarian_principle.

3. Sheldon Solomon, Jeff Greenberg, and Tom Pyszczynski, *The Worm at the Core: On the Role of Death in Life* (2015). See also the documentary film *Flight from Death: The Quest for Immortality*, by Patrick Shen and Greg Bennick (2006).

CHAPTER 5: GOING DOWN THE RABBIT HOLE

1. See, for example, the popular article by Moheb Costandi on the BBC website: "What Happens to Our Bodies after We Die" (2015).

2. Matthew Shaer, "Scientists Are Giving Dead Brains New Life. What Could Go Wrong?," *New York Times*, July 2, 2019; Max Kozlov, "What Does 'Brain Dead' Really Mean? The Battle over How Science Defines the End of Life," *Nature*, July 11, 2023; Sam Parnia and Josh Young, *Erasing Death* (2013).

3. Additional tools include, for example, nano-devices put inside the brain. See, for example, A. Paul Alivisatos et al., "Nanotools for Neuroscience and Brain Activity Mapping," *ACS Nano* 7 (2013): 1850–66.

4. See, for example, https://en.wikipedia.org/wiki/Hard_problem_of_consciousness. See also a recent review: Robert L. Kuhn, "A Landscape of Consciousness: Toward a Taxonomy of Explanations and Implications," *Progress in Biophysics and Molecular Biology* 190 (2024): 28–169.

CHAPTER 6: PREDICTABILITY AND A NEW DEFINITION FOR HUMAN LIFE AND DEATH

1. See also the simpler Turing-like motor-intelligence tests using a handshake: Ruth Stock-Homburg et al., "Evaluation of the Handshake Turing Test for Anthropomorphic Robots," *HRI '20: Companion of the 2020 ACM/IEEE International Conference of Human-Robot Interaction* (2020): 456, and references therein (original idea due to Amir Karniel of Ben-Gurion University of the Negev). See also https://en.wikipedia.org/wiki/Philosophy_of_artificial_intelligence.

2. There are numerous attempts to reverse engineer the brain by better computational algorithms running on standard silicon chips, or by also developing new hardware that may do a better job at emulating the brain. See, for example, Gert Cauwenberghs, "Reverse Engineering the Cognitive Brain," *PNAS* 110 (2013): 15512–13, https://doi.org/10.1073/pnas.1313114110. See also attempts to make a "brain transistor": X. Yan et al., "Moire Synaptic Transistor with Room-Temperature Neuromorphic Functionality," *Nature* 624 (2023): 551–56. Furthermore, see attempts to make the computer think like humans: Farshad Rafiei, Medha Shekhar, and Dobromir Rahnev, "The Neural Network RTNet Exhibits the Signatures of Human Perceptual Decision-Making," *Nature Human Behaviour* (2024), https://doi.org/10.1038/s41562-024-01914-8. Finally, see the textbook *Bayesian Models of Cognition: Reverse Engineering the Mind*, ed. Thomas L. Griffiths, Nick Chater, and Joshua B. Tenenbaum, The MIT Press (2024).

3. Aside from many academic papers on the topic, see also books by Ray Kurzweil, e.g., *The Singularity Is Near* (2005).

4. Peter Menzel and Faith D'Aluisio, *Robo Sapiens: Evolution of a New Species*, MIT Press (2001).

5. On advances in the interface between our brain and machines, see, for example, Ferris Jabr, "The Man Who Controls Computers with His Mind," *New York Times*, May 2022.

CHAPTER 7: CAN WE UNDO PREDICTABILITY?

1. Some Big Data scientists are already using the assumption of stochastic (random) processes to conjecture a theoretical upper bound on predictability. See, for example C. Krumme, A. Llorente, M. Cebrian, A. Pentland, and E. Moro, "The Predictability of Consumer Visitation Patterns," *Scientific Reports* 3 (2013): 1645.

2. A. Aldo Faisal, Luc P. J. Selen, and Daniel M. Wolpert, "Noise in the Nervous System," *Nature Reviews Neuroscience* 9 (2008): 292—303; Mark D. McDonnell, "The Benefits of Noise in Neural Systems: Bridging Theory and Experiment," *Nature Reviews Neuroscience* 12 (2011): 415–425.

3. Hua Tang and Bruno Averbeck, "Shared Mechanisms Mediate the Explore-Exploit Tradeoff in Macaques and Humans," *Neuron* 110 (2022): 1751–53.

4. See, for example, the work of Laura Carstensen from Stanford.

5. Yitzhak Norman and Rafael Malach, "What Can iEEG Inform Us about Mechanisms of Spontaneous Behavior?" See also Rafael Malach, "The Neuronal Basis for Human Creativity," *Frontiers in Human Neuroscience* 18 (2024): 1367922, https://doi.org/10.3389/fnhum.2024.1367922.

6. Yitzhak Norman and Rafael Malach, "What Can iEEG Inform us about Mechanisms of Spontaneous Behavior?" See also Rafael Malach, "The Neuronal Basis for Human Creativity," *Frontiers in Human Neuroscience* 18 (2024):1367922, https://doi.org/10.3389/fnhum.2024.1367922.

7. Y. Nir et al., "Interhemispheric Correlations of Slow Spontaneous Neuronal Fluctuations Revealed in Human Sensory Cortex," *Nature Neuroscience* 11 (2008): 1100–8.

8. Yitzhak Norman and Rafael Malach, "What Can iEEG Inform Us about Mechanisms of Spontaneous Behavior?," chapter in *Intracranial EEG for Cognitive Neuroscience* (2022).

9. Daniel Toker et al., "Consciousness Is Supported by Near-Critical Slow Cortical Electrodynamics," *PNAS* 119 (2022): e2024455119.

10. https://en.wikipedia.org/wiki/Neuronal_noise; http://www.scholarpedia.org/article/Neuronal_noise; http://www.scholarpedia.org/article/1/f_noise.

11. For example, E. Novikov et al., "Scale-Similar Activity in the Brain," *Physical Review E* 56 (1997): R2387–89, https://doi.org/10.1103/PhysRevE.56.R2387.

12. For example, L. M. Ward, *Dynamical Cognitive Science*, MIT Press (2002), describes an unpublished study by McDonald and Ward (1998).

13. Clement Moutard, Stanislas Dehaene, and Rafael Malach, "Spontaneous Fluctuations and Non-linear Ignitions: Two Dynamic Faces of Cortical Recurrent Loops," *Neuron* 88 (2015): 194–206.

14. Ariel Furstenberg, "From Reflex to Reflection: Moving from the Space of Causes to the Space of Reasons and Back," *Open Philosophy* 3 (2020): 681–93; A. Furstenberg, A. Breska, H. Sompolinsky, and L. Y. Deouell, "Evidence of Change of Intention in Picking Situations," *Journal of Cognitive Neuroscience* 27 (2015): 2133–46.

15. Max Tegmark, "Importance of Quantum Decoherence in Brain Processes," *Physical Review E* 61 (2000): 4194–206. Note that Tegmark calculates the decoherence of a superposition of millions of ions, which indeed is the standard picture of what happens when a neuron fires. He also calculates the decoherence rate of a mechanism suggested by Penrose in which quantum signals of polarization and kink-like excitations propagate through microtubules. He finds very short coherence times.

16. S. Hagen, S. R. Hameroff and J. A. Tuszynski, "Quantum computation in brain microtubules: decoherence and biological feasibility," *Physical Review E Stat. Nonlin. Soft Matter Phys.* 65 (2002): 061901. https://doi.org/10.1103/PhysRevE.65.061901

17. S. Kumar et al., *Scientific Reports* 6 (2016): 36508; P. Zarkeshian et al., *Scientific Reports* 12 (2022): 20720; Betony Adams and Francesco Petruccione, "Do Quantum Effects Play a Role in Consciousness?," *Physics Today*, January 2021.

18. Marlan Scully, plenary talk at the Photonics West Conference, San Francisco, January 2024.

19. See, for example, our recent publication and references therein: G. Amit, Y. Japha, T. Shushi, R. Folman, and E. Cohen, "Countering a Fundamental Law of Attraction with Quantum Wave-Packet Engineering," *Physical Review Research* 5 (2023): 013150.

20. For example, Carlo Rovelli, *Helgoland: Making Sense of the Quantum Revolution* (2021).

21. N. David Mermin, "Is the Moon There When Nobody Looks? Reality and the Quantum Theory," *Physics Today*, 1985.

22. Daniel M. Greenberger, in the book *Compendium of Quantum Physics* (2009), 258: "The GHZ states (Greenberger-Horne-Zeilinger states) are a set of entangled states that can be used to prove the GHZ theorem, which is a significant improvement over Bell's Theorem as a way to disprove the concept of 'elements of reality,' a concept introduced by EPR problem (Einstein-Podolsky-Rosen) in their attempt to prove that quantum theory is incomplete."

23. For entanglement between large objects, see Daniel Garisto, "Scientists Supersize Quantum Effects with Entangled Drum Duet," *Scientific American*, 2021.

24. See, for example, Jacob S. Higgins et al., "Photosynthesis Tunes Quantum-Mechanical Mixing of Electronic and Vibrational States to Steer Exciton Energy Transfer," *PNAS* 118 (2021): e2018240118. See also https://physicsworld.com/a/is-photosynthesis-quantum-ish.

25. *Quantum Effects in Biology*, ed. Masoud Mohseni, Yasser Omar, Gregory S. Engel, and Martin Plenio, Cambridge University Press (2014); Johnjoe McFadden and Jim Al-Kahlili, "The Origins of Quantum Biology," *Proceedings of the Royal Society A* 474 (2018): 20180674.

26. Betony Adams and Francesco Petruccione, "Quantum Effects in the Brain: A Review," *AVS Quantum Science* 2 (2020): 022901; "Do Quantum Effects Play a Role in Consciousness?," and references therein, https://physicsworld.com/a/do-quantum-effects-play-a-role-in-consciousness. See also Michel Gingras and Zoya Leonenko, "Searching for Quantum Effects in Neuroscience," *The Entangler*, University of Waterloo (2021). Also see works by Christoph Simon and colleagues: *Scientific Reports* 11 (2021): 6287; *Scientific Reports* 11 (2021): 12121; *Scientific Reports* 12 (2022): 269; *Scientific Reports* 12 (2022): 6109; *PLoS Computational Biology* 18 (2022): e1010198; *Journal of the Royal Society Interface* (2022).

27. See Luca Turin in Andrea Rinaldi, "Reawakening anaesthesia research," *EMBO reports* 15 (2014): 1113. https://www.embopress.org/doi/full/10.15252/embr.201439593.

28. See, for example, the work of the group of Michael Wasielewski at Northwestern University.

29. Mark Buchanan, "Nothing's Impossible," *Nature Physics* 7 (2011): 5.

30. Sean M. Carroll, "Physics and the Immortality of the Soul," *Scientific American*, Guest Blog (2011).

31. Buchanan, "Nothing's Impossible."

32. Y. Margalit et al., "Realization of a Complete Stern-Gerlach Interferometer: Towards a Test of Quantum Gravity," *Science Advances* 7 (2021): abg2879. See also Zack Zavitsky, "The (Often) Overlooked Experiment That Revealed the Quantum World," *Quanta* magazine, December 5, 2023.

33. See, for example, https://www.theguardian.com/science/blog/2015/oct/01/living-thing-two-places-limits-to-quantum-quandary.

34. See, for example, Jacob S. Higgins et al., "Photosynthesis Tunes Quantum-Mechanical Mixing of Electronic and Vibrational States to Steer Exciton Energy Transfer," *PNAS* 118, no. 11 (2021): e2018240118. See also https://physicsworld.com/a/is-photosynthesis-quantum-ish; *Quantum Effects in Biology*, ed. Masoud Mohseni, Yasser Omar, Gregory S. Engel, and Martin Plenio, Cambridge University Press (2014); Johnjoe McFadden and Jim Al-Kahlili, "The Origins of Quantum Biology," *Proceedings of the Royal Society A* 474 (2018): 20180674.

35. Stuart Hameroff and Roger Penrose, "Consciousness in the Universe: A Review of the 'Orch OR' Theory," *Physics of Life Reviews* 11 (2014): 39–78. See also Roger Penrose, *Shadows of the Mind: A Search for the Missing Science of Consciousness* (1994); Stuart Hameroff, "Quantum Computation in Brain Microtubules? The Penrose-Hameroff 'Orch OR' Model of Consciousness," *Philosophical Transactions of the Royal Society of London A* 356 (1998): 1869.

36. Julie Parato and Francesca Bartolini, "The Microtubule Cytoskeleton at the Synapse," *Neuroscience Letters* 753 (2021): 135850.

37. See, for example, the January 14, 2020, report of Jordana Cepelewicz in *Quanta* magazine, titled "Hidden Computational Power Found in Arms of Neurons," where she reports on Albert Gidon et al., "Dendritic Action Potentials and Computation in Human Layer 2/3 Cortical Neurons," *Science* 367 (2020): 83–87.

38. Nick E. Navromatos, "Quantum Coherence in (Brain) Microtubules and Efficient Energy and Information Transport," *Journal of Physics: Conference Series* 329 (2011): 012026; G. L. Celardo, M. Angeli, T. J. A. Craddock, and P. Kurian, "On the Existence of Superradiant Exitonic States in Microtubules," *New Journal of Physics* 21 (2019): 023005; K. Saxena et al., "Fractal, scale free electromagnetic resonance of a single brain extracted microtubule nanowire, a single tubulin protein and a single neuron," *Fractal Fract.* 4 (2020): 1; N. S. Babcock et al., "Ultraviolet

superradiance from mega-networks of tryptophan in biological architectures," *J. Phys. Chem. B* 128 (2024): 4035.

39. S. Kahn et al., "Microtubule-Stabilizer Epothilone B Delays Anesthetic-Induced Unconsciousness in Rats," *eNeuro* 11 (2024). https://doi.org/10.1523/ENEURO.0291-24.2024.

40. Jennifer Ouellette, "A New Spin on the Quantum Brain," *Quanta* magazine, November 2, 2016; Matthew P. A. Fisher, "Quantum Cognition: The Possibility of Processing with Nuclear Spins in the Brain," *Annals of Physics* 362 (2015): 593–602; "Are We Quantum Computers or Merely Clever Robots?," *International Journal of Modern Physics B* 31 (2017); M. W. Swift, M. P. A. Fisher, and C. G. Van de Walle, "Posner Molecules: From Atomic Structure to Nuclear Spins," *Physical Chemistry Chemical Physics* 20 (2018): 12373–80; S. Agarwal et al., "The Dynamical Ensemble of Posner Molecules Is Not Symmetric," *Journal of Physical Chemistry Letters* 12, no. 42 (2021): 10372. PS: The opinion of Matthew Fisher regarding the quantum mind and the Posner molecule is still favorable toward the end of 2023, when I had the pleasure of meeting him in person in Tokyo.

41. Lucien Hardy, "Proposal to Use Humans to Switch Settings in a Bell Experiment," ArXiv (2017), https://arxiv.org/pdf/1705.04620.

https://futurism.com/scientists-have-an-experiment-to-see-if-the-human-mind-is-bound-to-the-physical-world. For future directions in research into the quantum mind, see also Max Tegmark, "Consciousness as a State of Matter," *Chaos, Solitons & Fractals* 76 (2015): 238.

On recently funded research on the fundamental connection between quantum theory and the philosophy of the mind, see, for example, https://fqxi.org/community/articles/display/266.

42. See, for example, our own work with atomic magnetic and electric sensors: https://cordis.europa.eu/article/id/182748-quantumbased-neuroimaging-holds-the-promise-for-advanced-brain-mapping.

43. Na Li et al., "Nuclear Spin Attenuates the Anesthetic Potency of Xenon Isotopes in Mice: Implications for the Mechanisms of Anesthesia and Consciousness," *Anesthesiology* 129 (2018): 271–77. For the story of lithium, see Jennifer Ouellette, "A New Spin on the Quantum Brain," *Quanta* magazine, November 2, 2016. For the latest work from the Fisher group on lithium, see Aaron Ettenberg et al., "Differential Effects of Lithium Isotopes in a Ketamine-Induced Hyperactivity Model of Mania," *Pharmacology Biochemistry and Behavior* 190 (2020): 172875, https://doi.org/10.1016/j.pbb.2020.172875.

44. Betony Adams and Francesco Petruccione, "Quantum Effects in the Brain: A Review," *AVS Quantum Science* 2 (2020): 022901; "Do Quantum Effects Play a Role in Consciousness?," and references therein, https://physicsworld.com/a/do-quantum-effects-play-a-role-in-consciousness. See also Michel Gingras and Zoya Leonenko, "Searching for Quantum Effects in Neuroscience," *The Entangler*, University of Waterloo (2021). Also see works by Christoph Simon and colleagues: *Scientific Reports* 11 (2021): 6287; *Scientific Reports* 11 (2021): 12121; *Scientific Reports* 12 (2022): 269; *Scientific Reports* 12 (2022): 6109; *PLoS Computational Biology* 18 (2022): e1010198; *Journal of the Royal Society Interface* (2022).

45. A. Arena et al., "General Anesthesia Disrupts Complex Cortical Dynamics in Response to Intracranial Electrical Stimulation in Rats," *eNeuro* 8 (2021): 10.1523/ENEURO.0343-20.2021; Janisz Cukras and Joanna Sadlej, "Towards Quantum-Chemical Modeling of the Activity of Anesthetic Compounds," *International Journal of Molecular Sciences* 22 (2021): 9272. On the use of anesthesia to study consciousness, see also Zirui Huang et al., "Anterior Insula Regulates Brain Network Transitions That Gate Conscious Access," *Cell Reports* 35 (2021): 109081. See also a general review in *The Neurology of Consciousness*, ed. S. Laureys, O. Gosseries, and G. Tonini, second edition, Elsevier (2016).

46. Stuart Hameroff and Roger Penrose, "Consciousness in the Universe: A Review of the 'Orch OR' Theory," *Physics of Life Reviews* 11 (2014): 39–78. See also Roger Penrose, *Shadows of the Mind: A Search for the Missing Science of Consciousness* (1994); Stuart Hameroff, "Quantum Computation in Brain Microtubules? The Penrose-Hameroff 'Orch OR' Model of Consciousness," *Philosophical Transactions of the Royal Society of London A* 356 (1998): 1869.

47. See for example the Wikipedia entry "List of Unsolved Problems in Computer Science."

48. Robert Geroch and James Hartle, "Computability and Physical Theories," *Foundations of Physics* 16 (1986): 533–50.

49. Nicholas C. Stone and Nathan. W. C. Leigh, "A Statistical Solution to the Chaotic, Non-hierarchical Three-Body Problem," *Nature* 576 (2019): 406–10.

CHAPTER 9: THE PILL

1. For example, Yuval Noah Harari, *Sapiens: A Brief History of Humankind* (2015); David Graeber and David Wengrow, *The Dawn of Everything*, Penguin (2021).

2. Chemical balance in the brain motivating actions is more scientifically referred to as the reward system or pleasure centers, or hedonic hotspots. See, for example,

K. C. Berridge and M. L. Kringelbach, "Pleasure Systems in the Brain," *Neuron* 86, no. 3 (2015): 646–664; J. M. Richard, D. C. Castro, A. G. Difeliceantonio, M. J. Robinson, and K. C. Berridge, "Mapping Brain Circuits of Reward and Motivation," *Neuroscience & Biobehavioral Reviews* 37 (2013): 1919–1931; Daniel C. Castro and Kent C. Berridge, "Opioid and Orexin Hedonic Hotspots in Rat Orbitofrontal Cortex and Insula," *PNAS* 114, no. 43 (2017): E9215–34; M. L. Kringelbach and K. C. Berridge, "The Joyful Mind," *Scientific American* 307, no. 2 (2012): 44–45.

3. See, for example, Brittney, "What Motivates Artists to Create Art?," *Art Radar Journal*, November 16, 2021.

4. Jacob Goldenberg, *Inside the Box* (2013); Edward de Bono, *Six Thinking Hats* (1999). See recent scientific work using brain imaging: E. Bartoli et al., "Default Mode Network Electrophysiological Dynamics and Causal Role in Creative Thinking," *Brain* 147 (2024): 3409. Googling "innovation definition" returns more than a billion results.

5. See, for example, Keith J. Holyoak, *The Spider's Thread: Metaphor in Mind, Brain and Poetry*, MIT Press (2019); Patrick J. Kiger, "The Human Brain Is Hard Wired for Poetry," *How Stuff Works* (2017); Eugen Wassiliwizky et al., "The Emotional Power of Poetry: Neural Circuitry, Psychophysiology and Compositional Principles," *Social Cognitive and Affective Neuroscience* 12 (2017): 1229–40.

6. To get a glimpse of hidden predictability and predicting outliers, one may read a recent article in the *New York Times* about how China is already predicting the future behavior of outliers: "How China Is Policing the Future," *New York Times*, June 25, 2022.

EPILOGUE

1. Out of respect to the experience of mindfulness, I cite here one recent book on the subject: Sannon Harvey, *My Year of Living Mindfully* (2020).

2. Lene Rachel Andersen (Ed. Tomas Björkman), *The Nordic Secret: A European Story of Beauty and Freedom* (updated edition 2024).

APPENDIX A: THE BRAIN

1. There are so many books about the brain that it is hard to choose, but here are a few examples: Frank Amthor, *Neuroscience for Dummies*, 2nd edition (2016); Rita Carter, *The Human Brain Book* (2019); David Eagleman, *The Brain: The Story of You* (2017); Grace Lindsay, *Models of the Mind* (2021).

2. No claim is being made in this book that we understand how the brain works. I am claiming that one may come to interesting insights about the brain by measuring its predictability P.

PS: Science is just scratching the surface of understanding the brain. Thousands of scientists around the world are working hard, with numerous experimental tools and on several different levels, trying to understand how a network of neurons gives us cognition. Beautiful insights have been gained. One of my favorites is, "The brain is in the game of optimizing neuronal dynamics and connectivity to maximize the evidence for its model of the world"; Karl Friston, "Does Predictive Coding Have a Future?," *Nature Neuroscience* 21 (2020): 1019–21. To get just a tiny taste of recent work, see, for example, the January 14, 2020, report of Jordana Cepelewicz in *Quanta* magazine titled "Hidden Computational Power Found in Arms of Neurons," where she reports on Albert Gidon et al., "Dendritic Action Potentials and Computation in Human Layer 2/3 Cortical Neurons," *Science* 367 (2020): 83–87. As additional examples of the heroic and versatile work done on understanding the brain, one may look at these works: William Lotter et al., "A Neural Network Trained for Prediction Mimics Diverse Features of Biological Neurons and Perception," *Nature Machine Intelligence* 2 (2020): 210–19; David Zada et al., "Parp1 Promotes Sleep, Which Enhances DNA Repair in Neurons," *Molecular Cell* 81 (2021): 1–15; Goffredina Spano et al., "Dreaming with Hippocampal Damage," *eLife* 9 (2020): e56211. In addition, see Appendices A and B and many references throughout the book.

See also the European and US megaprojects with the goal of better understanding the brain: https://braininitiative.nih.gov; https://www.humanbrainproject.eu/en.

3. Adi Doron et al., Hippocampal astrocytes encode reward location, *Nature* 609 (2022): 772–778; R Refaeli, T Kreisel, TR Yaish, M Groysman, I Goshen, Astrocytes control recent and remote memory strength by affecting the recruitment of the CA1→ ACC projection to engrams, *Cell Reports* 43 (3) (2024): 113943.

4. Hans Helmut Kornhuber and Luder Deecke, *The Will and Its Brain: An Appraisal of Reasoned Free Will*, University Press of America (2012).

APPENDIX B: ATTEMPTS TO QUANTIFY CONSCIOUSNESS AND SELF-AWARENESS

1. *Explaining Consciousness: The Hard Problem*, ed. Jonathan Shear (1995).

2. J. E. LeDoux, M. Michel, and H. Lau, "A Little History Goes a Long Way toward Understanding Why We Study Consciousness the Way We Do Today," *PNAS* 117 (2020): 6976–84. And here is the original Crick and Koch paper: Francis Crick and Christof Koch, "Towards a Neurobiological Theory of Consciousness," *Seminars in*

Neurosciences 2 (1990): 263–75. PS: I must say I am quite amazed that in October 2023, only 3,091 published works cite this paper.

3. Andrea Nani et al., "The Neural Correlates of Consciousness and Attention: Two Sister Processes of the Brain," *Frontiers in Neuroscience* (2019), https://doi.org/10.3389/fnins.2019.01169.

4. There are many references throughout this book, and here I add additional ones: Anil Seth, *30-Second Brain* (2014); Anil Seth, *Being You* (2021); Daniel. J. Levitin, *This Is Your Brain on Music* (2008); and the more academic reviews: David L. Barack and John W. Krakauer, "Two Views on the Cognitive Brain," *Nature Reviews Neuroscience* 22 (2021): 359–71; *The Neurology of Consciousness*, ed. S. Laureys, O. Gosseries, and G. Tonini, 2nd edition, Elsevier (2016); Miguel Nicolelis, *Beyond Boundaries: The New Neuroscience of Connecting—and How It Will Change Our Lives* (2012). There are also numerous, perhaps dozens, less-know attempts to understand consciousness, for example: N. Lahav and Z. A. Neemeh, "A Relativistic Theory of Consciousness," *Frontiers in Psychology* 12 (2022): 704270.

5. An alternative division may be found on p. 414 (chap. 25) in *The Neurology of Consciousness*, ed. S. Laureys, O. Gosseries, and G. Tonini, 2nd edition, Elsevier (2016): autobiographical self, feeling of agency, feeling of ownership.

6. Special issue of *Neuron* on consciousness, vol. 112, no. 10 (May 15, 2024): 1519–722.

7. Q & A with Stanislas Dehaene (2024): https://www.cell.com/neuron/fulltext/S0896-6273(24)00159-4.

8. Takuya Niikawa, "A Map of Consciousness Studies: Questions and Approaches," *Frontiers in Psychology*, October 8, 2020. Robert L. Kuhn, "A Landscape of Consciousness: Toward a Taxonomy of Explanations and Implications," *Progress in Biophysics and Molecular Biology* 190 (2024): 28-169.

9. Giulio Tonini, "An Information Integration Theory of Consciousness," *BMC Neuroscience* 5 (2004): 42.

10. Rufin VanRullen and Ryota Kanai, "Deep Learning and the Global Workspace Theory," *Trends in Neurosciences* 44 (2021): 692–704, https://arxiv.org/pdf/2012.10390.pdf.
See also https://en.wikipedia.org/wiki/Global_workspace_theory.

11. G. A. Mashour, P. Roelfsema, J.-P. Changeux, and S. Dehaene, "Conscious Processing and the Global Neuronal Workspace Hypothesis," *Neuron* 105 (2020): 776–98.

NOTES

12. Jason Pontin, "How Do You Know You Are Reading This?," *Wired*, April 2, 2018, https://www.wired.com/story/tricky-business-of-measuring-consciousness; Emily Sohn, "Decoding the Neuroscience of Consciousness," *Nature* 571 (July 24, 2019): S2–S5. Concerning IIT, see also http://www.scholarpedia.org/article/Integrated_information_theory.

13. Karim Jerbi and Jordan O'Byrne: https://fqxi.org/community/articles/display/265.

14. https://osf.io/preprints/psyarxiv/zsr78.

15. Lucia Melloni, Liad Mudrik, Michael Pitts, and Christof Koch, "Making the Hard Problem of Consciousness Easier," *Science* 372 (2021): 911–12. See also https://www.youtube.com/watch?v=j0H5em2tMbs.

16. Stanislas Dehaene and Jean-Pierre Changeux, "Experimental and Theoretical Approaches to Conscious Processing," *Neuron* 70 (2011): 200–227.

17. Daniel Revish and Moti Salti, "Expanding the Discussion: Revision of the Fundamental Assumptions Framing the Study of the Neural Correlates of Consciousness," *Consciousness and Cognition* 96 (2021): 103229. See also Adrien Doerig et al., "The Unfolding Argument: Why IIT and Other Causal Structure Theories Cannot Explain Consciousness," *Consciousness and Cognition* 72 (2019): 49–59; M. H. Herzog et al, "First-Person Experience Cannot Rescue Causal Structure Theories from the Unfolding Argument," *Consciousness and Cognition* 98 (2022): 103261.

18. Matthias Michel et al., "Opportunities and Challenges for a Maturing Science of Consciousness," *Nature Human Behavior* 3 (2019): 104–7.

APPENDIX C: RESEARCH DEALING WITH PREDICTION OF HUMAN BEHAVIOR

1. https://futureoflife.org/ai/six-month-letter-expires; https://sifted.eu/articles/mistral-aleph-alpha-and-big-techs-lobbying-on-ai-safety-will-hurt-startups; https:/futureoflife.org/ai-policy/protect-the-eu-ai-act; https:/artificialintelligenceact.eu. See also statement by Max Tegmark to Congress: https://futureoflife.org/ai-policy/written-statement-of-dr-max-tegmark-to-the-ai-insight-forum.

2. https://www.wilmerhale.com/en/insights/blogs/wilmerhale-privacy-and-cybersecurity-law/20231219-the-eu-reaches-a-political-agreement-on-the-ai-act.

3. See for example the advantages of a multidimensional data set taken of an individual in enabling high-quality predictions: Wanyi Zhang et al., "Putting

Human Behavior Predictability in Context," *EPJ Data Science* (2021), https://doi.org/10.1140/epjds/s13688-021-00299-2; See also the emerging field of context awareness, e.g., https://en.wikipedia.org/wiki/Context_awareness.

4. https://www.gottman.com/blog/the-research-predicting-divorce-from-an-oral-history-interview.

5. Jonathan Stray, Alon Halevy, et al., "Building Human Values into Recommender Systems: An Interdisciplinary Synthesis," arXiv (2022), https://arxiv.org/pdf/2207.10192.

6. One of the concluding statements of this PhD thesis by Gal Ben Yosef (2019) is, "The predictive approach to personality modeling could theoretically lead to models that render human behavior extremely predictable."

7. https://cos.northeastern.edu/news/human-behavior-is-93-predictable-research-shows.

Obviously there is a huge gap between the popular declarations and the detailed scientific articles, in this case *Science* 327, no. 5968 (2010): 1018–21, in which one finds that this work focused on mobility, and that for individuals they could only reach 80 percent predictive accuracy.

Index

action potential, 218. *See also* brain outcomes
advertising, East German Stasi and, 36
aesthetics: kitsch and, 233–34; pleasure and, 194–97
AI. *See* artificial intelligence
AI and Big Data, 8, 19; DNA test scenario and, 34; extinction and, 10; happiness and contentment provided by, 193–94; political dissents and, 189–90; predictive technology using, 27–31. *See also* Big Data
algorithms, 229; chemical, 72; deep-learning, 41–43, 231; platform economy and, 227
"am I really alive" question, xi–xii, xiv, 44, 211
Anderson, Laurie, 99
Anderson, Philip, 62
Angell, Ian, 182
angstroms, 150
Animals, 223

animal welfare bill, 223
answer, to Eve's question, 19–23
antimatter, 156
Ariely, Dan, 49
Aristotle, 62
artificial intelligence (AI): AI Act in EU on, 227; anxiety over, xv–xvi; art competition won by, 14; context awareness of, 42; human predictability proof by, xvii; introspection afforded by, ix; predictive technologies based in, 15–16; P score calculation by, 37; source of computing power of, 36–37; target function in, 165; value alignment, xvi, 41, 231. *See also* AI and Big Data
artists, 195–96; revolt assignment for, 200–201
Ashkin, Arthur, 101
atomic bomb, 16
atomic clocks experiment, 105, 106–8, 152

atoms: are they alive question, 2–3, 19, 110, 153–54, 211; atomic dark matter detector, 6; atomic hypothesis, 61–62; brain as, 4, 168; communication with, 1–2, 8, 17; discrete feature of, 145–46; electron firing number of, 143; importance of, 61–62; ionization, 100; light interaction of, 3; noise and, 150; vacuum chamber isolation of, 150
Auschwitz death camp, ix–x; letter requesting women from, 118
Austerlitz (Sebald), 44–45
aware person, 215. *See also* self-awareness
Aztec sacrifices, 188

The Battle for Your Brain (Farahany), 228
BCI. *See* brain-computer interface
Becker, Ernest, 94
beetle in box (thought experiment), 122, 224
Behavio, 38, 229
behavior, human: chemical basis of, 54; prediction research, 227–31, 251n7; universal models of, 41–42
Belinson, Moshe, 4
Bell, Gordon, 40
Bell, John, 79, 154
bell curve, 207
Bell Labs, 30–31
Bereitschaftspotential (BP), 137
Bernays, Edward, 37–38, 190
beyond-man, Nietzsche idea of, 215
Bezalel Academy of Arts and Design, 200
Big Data, xvi–xvii, 39, 228–29

binary star, 104
blackbody radiation, 145, 150
Black Mirror, 125
Blake, William, 91
block universe, 50
Bohr, Niels, 146–47
BOLD imaging, 141
"born," 4
BP. *See Bereitschaftspotential*
Bradley, A. C., 205
brain: altering predictability of, 16; as atoms, 4, 168; coherence, 144; collapse and decoherence theories, 157–60; consciousness and, 86, 212; cortex, 140–41, 219; creativity and, 202; current knowledge about, 45–46; data harvesting and access to, 228; differences between predictability according to, 15, 175; entanglement applied to, 149–50, 159, 173–74; filters, 165–67, 172, 201; future instruments for understanding, 123–24; in glass vessel, 109–11; implants, 40; injuries, 225; input quality for deductions about, 123–24; kuru disease of, 120; limited knowledge of, 109; new physics for understanding, 154–62; number of neurons comprising, 219; output, 163; parts of, 219; poetry and, 204; quantum field theory (QFT) and, 51–52, 143, 173–74, 238n14; randomness and, 35; resting state of cortical networks, 140–41; strangeness of QFT and, 148–50; supercomputer simulation of, 148–49; superradiance effect, 144; synapse, 159, 217; transcranial magnetic stimulation of, 225;

two-layer model of, 35; unique behavior of each, 56; varying levels of predictability in, 57; will and, 219. *See also* decoherence; noise, internal brain

brain, determinism question for, 44; imaginary supercomputer analysis of, 46–48, 53, 55; Laplace demon and, 49–53, 56; neurons in predicting, 53, 55; P score as able to settle, 57

brain-computer interface (BCI), 40, 228

brain outcomes (actions): predictability and, 7; quantum processes impact on, 143; two-layer model for, 163–68, 174

brainwashing, 55–56

Broglie, Louis de, 70, 101, 146

Burroughs, William S., 99

Cambridge Analytical data, Facebook scandal, 11–12

Camus, Albert, 216

cannibalism, 120

Carroll, Sean M., 78–79

Cavendish, Henry, 108

CERN, 47, 66

Challenger, 191

chaos theory, 53, 153

chef analogy, 123

chemical balance, 192–93, 199, 240n6; aesthetic pleasure and, 194–95; evolution and, 198; making art to achieve, 196; of pleasing calm, 197; predictability suppressed by, 196–97

chemistry: chemical algorithms, 72; consciousness and, 54–55; of love, 54

China, 186

cigarettes, as "torches of freedom," 37

Clifford, Will, 106–7

CMT. *See* computational theory of mind

Cohen-Tannoudji, Claude, 3, 19, 154

coherence, brain, 144

collapse postulate, 147–48; brain relevance of decoherence and, 157–60; consciousness and, 158

colored noise, 139

comfort zone, 87–88; new physics and, 155; pleasing calm, 198

common sense, Einstein on, 88

compatibilism, 65

computational theory of mind (CTM), 125

Confucius, 3, 22, 110–11; on second human life, 118–19, 179

consciousness: atoms and, 48; attempts to quantify, 221–23, 225–26; brain and, 86, 212; chemical basis of, 54–55; collapse caused by, 158; definition lacking for, 18; easy problems of, 221; experimental search for, 6–8; flawed definitions based on, 128; "hard problem of," 121–22, 221, 222–23; "higher," 14–15; ITT attempts to identify properties of, 225–26; mind-altering drugs causing higher, 13, 14, 131; other terms for, 122; personal growth and "natural," 215; popular theories of, 8; predictability connection to, 12–15; predictability P basis of definition, 212; present theories of, 212; P value and brain states correlation with, 175–76; science and, 168; study of, xiv; Wikipedia on, 121

context awareness, of AI, 42

continuous spontaneous localization (CSL), 157
Conway, John, 73
Copernicus, Nicolaus, 92, 93, 96
cortex, 219; sensory, 140–41
creativity: brain random fluctuations and, 202; fighting predictability with, 199–202; free will and, 74, 138; predictability as limiting, 13, 14, 15; society benefit from, 208
Crick, Francis, 222
CSL. *See* continuous spontaneous localization
CTRL-labs, 228
culture: death denial in, 94, 95; fear of death in collective, 96; of science and technology versus humanities, 183
cyborgs, 98, 127–28; definition of, 40

Dada movement, 183, 187–88, 216
dark energy, 156
dark matter detector, 6
Darwin, Charles, 86; ego blow inflicted by, 92–94, 96; son of, 215
Data (robot in *Star Trek*), 125–26
data harvesting, xvi–xvii, 43; by dark agents, 39; political use of, 11; public declarations about, 41; regulation issue, 227; "secure" storage of data, 41; smart homes and, 228; by social media, 39; trends to cause increased, 40; wearable devices for, 228
death: as archenemy of ego, 92; Auschwitz death camp, ix–x, 118; body decay after, 120; culture based on denying, 94, 95; denial of, 92–97; as insult to spirit, 94–95; mental, 18, 20, 45, 97; need for redefining life and, 87, 120–21; predictability as, 131; questionnaires, 95–96; redefining, 17, 87; redefining life and, 87, 120–21; redefining life by defining, 129–30. *See also* machine death
decision-making processes: entanglement and, 174; exploit-explore trade-off in, 137, 180; filters in, 165–67, 172; free behavior and fluctuations implications for, 141; Gaussian function for analyzing, 207; as human feedback loops, 71–72; quantum computers modeling, 152
decoherence: brain in quantum collapse and, 157–60; cause of, 150–52; entanglement and, 149–50, 159; measurement, 143, 243n15; rates, 159–60; suppression, 151–52
Deecke, Lüder, 74
deep-learning machines, 41–43, 230, 231
degrees of freedom, 109
denial-of-death theory, 94–96
Dennett, Daniel, xii
Descartes, Rene, 3, 118, 126; "I doubt" of, 127, 199
Determined (Sapolsky), xvii, 66
determinism: definition of, 20; feedback loops and, 71–72; filters as responsible for, 166, 172; free energy principle and, 8; free will compatibility with, 65; Freud and, 69–70; internal, 186; irrationality argument against, 49; jail of, 199; Laplace demon and, 49–53, 56; mind and, 46; noise as part of, 142; proponents of, 65, 66; of quantum mechanics, 161–62; quantum

randomness differing from strong, 173; question on brain and, 44, 45–46, 48; randomness *vs.*, 23; strong and weak, 171–73
"deterministic," xvii
diffraction (interference), 99–100
digital footprint, xvii, 34; "intelligent histories," 39
digital phenotyping, 38, 230
Dingle, Herbert, 106
divergent thinking, 199–200
divorce, predictability rate for, 229–30
DNA, 16, 33–34
dopamine, 54
Doppler effect, 108
double-slit experiment, 147, 158
drugs, mind-altering, 87, 198; higher consciousness from, 13, 14, 131; poetry as, 205. *See also* pill, anti-predictability
duality, language and, 100–101
Dylan, Bob, 13

Eastern religions, 65
East German Stasi, 36
Ecclesiastes, 216
Eddington, Arthur, 108
EEG, 175, 204; brain potential discovered through, 137–38
ego: Darwin theory as blow to, 92–94, 96; death as archenemy of, 92; fighting, 92–97
Einstein, Albert, 23, 47, 70, 93; Bohr and, 147; on common sense, 88; on logic, 203; machine death theory and, 111; Newton and, 104; quantum randomness hated by, 146; reactions to theory of, 105–6; relativity and, 50; relativity theory of, 105–6, 154; special and general theories of relativity, 107; theory of invariance, 105; thought experiment of, 103, 107
Eisenhower, Dwight, 188–89
electrical circuits, noise in, 139–40
electric shocks, 190, 192
electromagnetic spectrum, partial visibility of, 98
electrons: discoverer of, 101; noise and, 140; thermal velocity of, 140
emergence: definition, 63; field of, 5; reductionism versus, 61–64
emotions, predictability of, 43
The Emperor's New Mind (Penrose), 125
entanglement, 149–50, 159; in brain, 173–74
Enterprise (spaceship), 125–26
Entropy, 152
epistemological theory, 224
EPJ Data Science, 228
erobotics, 228
EU AI Act, 227
eugenics movement, xiii, 215
Eve (filmmaker): are they alive question by, 2–3, 19, 110, 153–54, 211; final words during lab visit, 21–22; imaginary discussion with, 5–6, 18–19; imagining more questions from, 5–6, 10; intricate request of, 181; lab visit by, 2–4; question asked by, 2–3, 211; request to virtual, 216
evolution, 22, 198; language and, 99
existential gap, 29, 183–84, 185
experiment, on predictability, 23; definition, 27–28; "free will" absence

for P1 success and, 43–44; human machine and, 44–57; imaginary zombie (Nikon camera), 31–32; ongoing, 168; under our radar, 43; purposes, 27; secrecy of, 44
experiments: atomic clocks, 105, 106–8, 152; consciousness research, 121; difficult interpretations of, 86–88; double-slit, 147, 158; electric shock (Milgram), 190; inadvertent, 27, 30–31, 36; interpreting results of, 116; Libet-type, 164; naturally-occurring, 30; reliability of well-performed, 27–31; on women (IG Farben), 118. *See also* thought experiment
exploit-explore trade-off, 137, 180
extinction, 9, 10, 131

Facebook, data scandal on, 11–12
Falkenburg, Brigitte, 63
FAPP. *See* for all practical purposes
Farahany, Nita, 228
feedback loops, 71–72, 141
Fermi paradox, 29–30
Feynman, Richard, 45, 147, 191; on atoms importance, 61–62
filters, 165–67, 172, 201
First World War, 216
Fisher, Matthew, 159
fMRI, 175, 204
for all practical purposes (FAPP), 136, 142, 170; P value and, 171–72
Fore people, kuru brain disease among, 120
The Forty Days of Musa Dagh, 187
Fourier transform, 139
Franck, James, 145–46
Free Agents (Mitchell), 66

free behavior (free thought), 138; for all practical purposes (FAPP), 136, 142, 170; fluctuations basis of, 141; noise as source of, 164
freedom: acquiring, 172; noise and, 138
free energy principle, 8
free will, 135, 219; claim that survival is proof of, 70–71; compatibility with determinism, 65; complexity as limiting, 48, 75–76; creativity and, 74, 138; demotion, 12; karma and, 65; as nondeterministic, 49; P1 requirement of total absence of, 43–44; predictability precluding, 9, 11, 12, 174; recent approaches to question of, 73–74; soldiers and, 186
free won't, 73, 164
Freud, Sigmund, 44, 87, 92, 164; Bernays and, 37; death denial and, 94; deterministic mind promoted by, 69–70; ego blow inflicted by, 96; iceberg analogy of, 222
Friston, Karl, 8, 55
Frontiers of Psychology, 223–24
Future of Humanity Institute, 9
Future of Life Institute, xv–xvi

GABA. *See* gamma-aminobutyric acid
Galileo Galilei, 93, 207
gaming industry, random number generators in, 142–43
gamma-aminobutyric acid (GABA), 218
Gaussian function, 207
GCS. *See* Glasgow Coma Scale
GDPR. *See* General Data Protection Regulation
Gell-Mann, Murray, 90
general anesthesia, 54, 151

General Data Protection Regulation (GDPR), 227
generalist-specialist (GS) score, 75
genocide, xiii, 186, 187
Geroch, Robert, 161
Ghiradi-Rimini-Weber (GRW) model, 157
Glasgow Coma Scale (GCS), 14
glass vessel, brain in, 109–11
global neural workspace theory (GNW), xiv, 55, 121, 224, 225
God, belief in, 127
Goebbels, Joseph, 38
Goff, Philip, 161
Goldfaden, Abraham, 5
Gordian knot, 181
Gottman, John, 38, 229–30
GPS, 105
gravity, 157
Greenberg, Jeff, 94
Greenpeace, x
GRW. *See* Ghiradi-Rimini-Weber
GS. *See* generalist-specialist
Guatemala, Bernays propaganda in, 38

Haiku, 205
half-life of facts, 28
Hänsch, Theodor (Ted), 155
happiness: aesthetics and, 194–96; AI and Big Data providing widespread, 193–94; predictability and, 192–94; redefining, 214; skepticism as condition for, 197; types, 199
Harari, Yuval Noah, xviii
"hard problem of consciousness," 121–22
Hartle, James, 161
Hawking, Stephen, 63

Haynes, John-Dylan, 73
Hegel, Georg Wilhelm Friedrich, 215
Heisenberg, Werner, 146
Hendrix, Jimi, 13
Hertz, Gustav, 145–46
HEXACO, 230
hidden predictability, 208–9
history, genocides not taught in, 187
Hitchhiker's Guide to the Galaxy, 97
Hobbes, Thomas, 207
Hofmann, Albert, 13
Holocaust, 187
Homo Deus (Harari), xviii
hospital monitors, x, xiii
humanities, 168; culture of science versus culture of, 183; logic versus, 182; request for answer in language of, 181
humans: AI value alignment with, xvi, 41, 231; as living machines, 174; name for future, 214–15; robots and, 124–26, 128; truth as hard for, 89. *See also* behavior, human; predictability
Hume, David, 194–95
Hundertwasser, Friedensreich, 198

iceberg analogy, of Freud, 222
iEEG. *See* intracranial EEG
IG Farben, 118, 183
IIT. *See* integrated information theory
imagination, 203, 206, 214
immortality, 94, 95
indoctrination, 186
integrated information theory (IIT), xiv, 8, 15, 224, 226; free will and, 74; inventor of, 225
intelligent history, 37, 39, 43, 169; brain determinism and, 56

The International Dictionary of Psychology, 221
intracranial EEG (iEEG), 40, 138
introspection, AI and, ix
ionization, atomic, 100
IonQ, 152
irrationality, 49, 80

Jobs, Steve, 13
John Deere, 227
Johnson-Nyquist noise, 140
Journal of Consciousness Studies, 221
judges, death question asked of, 95–96

Kafka, Franz, 216
Kahneman, Daniel, 10
Kant, Immanuel, 92, 164, 195; on soul, 76; on truth distortion, 89
karma, 65
Kauffman, Stuart, 198
Keen, Sam, 94–95
Kepler, Johannes, 101
King, Martin Luther, Jr., 189
kitsch, 233–34
Koch, Christof, 222
Kochen, Simon, 73
Kornhuber, Hans Helmut, 74
Kuhn, Thomas, 98
Kulka, Tomas, 233
Kundera, Milan, 233
Kurzweil, Ray, xv, 79

Laboratory for National Eugenics, 215
language, 181; duality and, 100–101; evolution and, 99; limits of, 98–102
Laplace, Pierre-Simon, 49–50
Laplace demon, 49–53, 56, 67, 191; quantum randomness not predicted by, 142; weak determinism in spite of, 172
Large Hadron Collider, 47
law of inertia, 67–68
laws of nature: decoherence rates and, 160; free will and, 67–68; inventing mental versions of, 68–69; reductionism and, 63; seed of unpredictability based on, 175; totalitarian principle and, 90
Leibniz, Gottfried Wilhelm, 204
Lenard, Philipp, 106
Leopold the Builder, 187
libertas disputandi, 155, 194
Libet, Benjamin, 73
Libet-type experiments, 164
Lichtenberg, 90
life, definition of: *consciousness* as flawed basis for, 128; death definitions and, 129–30; freedom and, 172; measuring amount of, 132; middle position of, 143; NASA and, 3, 4, 117; philosophical pause in, 20; preparing for criticism of new, 105; redefining, 128–31; redefining death and, 87, 120–21; robots and humans comparison for, 124–26, 128; Sartre and, 124; search for new, 22; second human, 118–19, 179; UBNR as new, 180; unpredictable but not random (UBNR), 130, 131. *See also* unpredictable but not random
light: bending, 108; cone, 47; speed of, 104; wave and particle properties of, 99, 100
"List of Unsolved Problems in Physics," 156–57, 244n21

logic, 182, 183, 208, 213; suppressing, 202–6
longevity, biological, 17, 117
lower-level sciences (mechanisms), 63
LSD, 13

MacDougall, Duncan, 78
machine breaking, 216
machine death, 8–12; ego and, 94; Einstein and, 111; new P observable implications for, 169; P value for, 44, 171–72; soul and, 77–78
machine learning, 41–43, 230, 231
machines, fundamental definition of, 129
Magritte, René, 102, 181, 205
Maimonides, 65
many-worlds interpretation, 90, 157
Marxism, 70
Massimini, Marcello, 225
The Matrix, 11
measure, 7
measurement: amount of life, 132; decoherence, 143, 243n15; future scenario of DNA test, 33–34; listening diversity in Spotify, 75; measurable observable, 37, 64; noise, 139–42; objectivity of instruments for, 87; observer affecting, 149; predictability, 8, 19, 31–44; of randomness, 130–31; self-awareness and consciousness, 221; unrealistic example (Nikon camera), 31–32
MEG, 175; measuring noise, 140
mental death, 18, 20, 45; ego hurt by idea of, 97
Mental Deficiency Act, 215
mental inertia, 68–69

mental laws, 68
metaphysics: psychology as based on, 53–55; soul notion in, 76
Midjourney, xv, 14
Milgram experiments, 190
military industrial complex, 188–89, 191
mind, computational theory of, 125
mind-body problem, xiii, 5, 76
Mind Strong, 230
mirror test, 223
missionaries, 120
Mitchell, Kevin, 66
MIT Media Lab, 38, 229
models, phenomenological, 19
mold, 117
Molecular Psychiatry, 54
molecules, water, 62
monkeys, self-awareness in, 223
moral responsibility, 65–66
"More Is Different" (Anderson), 62
Morrison, Jim, 91
Morrison, Margaret, 63
mountain peak analogy, 212
Musolino, Julien, 79

Namibia, genocide in, xiii, 215
NASA, 191; life definition and, 3, 4, 117
Nazis, 38, 186, 215
neural correlates of consciousness (NCC), 222
Neuron, 223, 226
neurons: action potential and, 218; brain analysis through, 53, 55; definition of, 217; excitatory and inhibitory, 218; number of brain, 219
neurotransmitters, 54, 217–18
New Guinea, Fore in Papua, 120

new physics, for understanding brain: consciousness open question and, 161; "List of Unsolved Problems in Physics," 156–57; noncomputable systems and, 161–62; open questions and, 160–61; quantum mind and, 51, 157–60; science evolution and, 154–55; source of, 156
Newton, Isaac, 104
Newtonian dynamics, 93
Newton's first law, 67
Nietzsche, 215
noise, internal brain, 163–64, 180; blackbody radiation and, 150; definition of, 136; determinism and, 142; freedom and, 138; measurement and, 139–42; predictability undone by, 135–39; quantum mind and, 142–54; resting-state fluctuations and, 140–41; two-layer model and, 166; unknown source of, 137

objectivity, 102; language limits and, 98–102; of measuring instruments, 87; quantum measurement and, 149; truth and, 88–92
observables, 12; predictability as measurable, 36–37
Oedipus complex, 69–70
ontology, 224

Papua, New Guinea, Fore in, 120
particle accelerators, 104, 145
particles, as also waves, 101
Pascal, Blaise, 127, 172, 202
PCI. *See* perturbational complexity index
PEA. *See* phenethylamine

Penrose, Roger, 51, 125, 143; decoherence rates and, 158; on new physics, 161
Penzias, Arno, 30–31
personality traits: filters as, 165–66; predictability as, 170
perturbational complexity index (PCI), 225
Petrov, Stanislav, 189
phenethylamine (PEA), 54
philosophy: meaning of word, 85; science as secondary to, 183; science versus, 85, 182–83
photoelectric effect, 100, 146
photosynthesis process, 151
physics: poetry as above, 205; social, 207; unanswered questions in, 156
Picasso, Pablo, 13
pill, anti-predictability (anti-zombie), 15–17, 21, 43, 176; creativity ingredient in, 199–202; defining what *ought to be* for, 185; emancipation and, 191; first ingredient, 192; of first kind, 182, 199, 213; happiness and, 192–99; logic suppression in, 202–6; poetry and, 204–6; P value and, 184; science-humanities gap and, 183–85; of second kind, 182, 213; society and, 206–9; war and, 186–91
Planck, Max, 146; ultraviolet catastrophe solved by, 145
platform economy, 227–28
pleasure, aesthetics and, 194–97
poetry, 204–6
"Poetry for Poetry's Sake" (Bradley), 205
political dissents, 189–90
Popper, Karl, 70; on search for truth, 89
Posner molecule, 159

Powell, Colin, 189
predictability: as absence of free will, 174; academic focus on, 41; aesthetics and, 194–96, 233–34; AI as proving human, xvii; brain outcomes and, 7; chemical balance as suppressing, 196–97; consciousness connection to, 12–15; creativity lessened by, 13, 14, 15; cross-discipline, 229; current uses of, 38–40; as death, 131; decoherence and, 150–52; different brains showing varying, 163, 167–68; of emotions, 43; enhancers, 190–91, 212–13; forces driving field of, 10–11; future tests for, 176; happiness and, 192–94; hidden, 208–9; implications of high level, 85; indoctrination and, 186–87; of intricate thoughts, 43; machine death and, 8–12; of machines, 129; many aspects to, 19; normalized, 171; obedience and, 190; percentage for behavior, 231; as personality trait, 170; poetry as suppressing, 205; potential uses of, 37–38; proof, 229; public versus science data on, 251n7; redefining happiness for fighting, 214; research on behavior, 227–31, 251n7; resonance and kitsch in understanding, 233–34; science and, 21, 135; skepticism absence in, 188–89; social pyramid of, 190, 208–9; "tribes" and, 41–42; unpredictable but not random (UBNR) and, 130, 131; users and exploitation of, 11–12; varying levels of brain, 57; war examples of, 186–91. *See also* free will; pill, anti-predictability; P value; unpredictability; unpredictable but not random
predictability, measuring, 8, 10, 31–44; as already underway, 19
predictability, undoing: brain noise and, 135–39; noise measurement and, 139–42; quantum randomness and, 152–53. *See also* P value
Predictably Irrational (Ariely), 49
predictive technology, using AI and Big Data, 27–31
Proctor, Helen S., 223
psychology, metaphysics and, 53–55
punishment, 167
Pure Matching, 230
Pushed to the Edge, 190
P value (P score): V and, 35; AI calculation of, 37; brain outcomes measuring individual, 169; brain states and consciousness correlation with, 175–76; bran determinism settlement by, 57; defining what *ought to be* with, 185; definition, 34–35; definition of life based on, 180; in different brains, 175; FAPP and, 171–72; "free will" absence for P1, 43–44; for machine death, 44, 171–72; middle ground (P = 0.5), 212; mind-altering drugs P=1, 200; P = 0, 35; soul test and, 80–81; for UBNR, 169
Pyszczynski, Tom, 94

QFT. *See* quantum field theory
Quanta, 4
quantum computers, 36–37; decision-making modeling by, 152

quantum field theory (QFT), 23; biology and, 151; birth of quantum mechanics, 145–48, 156; brain and, 51–52, 143, 173–74, 238n14; brain determinism and, 46–48; collapse postulate added to, 147–48; determinism of, 161–62; double-slit paradigm in, 147, 158; electron discovery, 101; entanglement, 149–50, 159, 173–74; many-worlds interpretation of, 90, 157; Nobel Prizes for, 146; quantum circuits, 152; quantum eraser, 152; randomness (statistical theory) in, 51, 142–43, 144, 173–74; soul and, 78–79; strangeness of, 148–50; superposition principle, 146–47; Tegmark calculations in, 143, 243n15; ultraviolet catastrophe and, 145; unanswered questions in, 156–57, 244n21; uncertainty principle, 50–51, 52, 148

quantum information mediators, 144

quantum mechanics. *See* quantum field theory

quantum mind, 51, 157–60; decoherence and, 143, 243n15; randomness and, 142–43

quantum science laboratory: communication with atoms in, 1–2; Eve visit to, 2–4

questions: "am I really alive," xi–xii, xiv, 44, 211; of Eve (are they alive), 2–3, 19, 110, 153–54, 211; five parts to answer, 19–23; free will of soldiers, 186; Goldfaden "who am I," 5; imagined, 5–6, 10, 12; QFT unanswered, 156–57, 244n21; for thought experiment, 115–19. *See also* brain, determinism question for

randomness, 182, 202; brain and, 35; noise as, 136; P = 0 and, 35; as part of quantum theory, 142–43, 144, 173; predictability undoing by quantum, 152–53; quantum, 51, 142–43, 144, 173; in two-layer model, 166–67. *See also* unpredictable but not random

random number generators, 142–43, 164

readiness potential (RP), 73, 137–38; filters causing delay in, 165

recommender systems, 10, 39

red shift, 107

reductionism: emergence versus, 61–64; laws of nature and, 63

Reis, Ben, 38

relative freedom, 138

relativity theory, 105–6, 154; source of, 156

resonance, predictability and, 233–34

resting-state fluctuations, 140–41

robots: comparing humans with, 124–26, 128; erobotics, 228; robo-sapiens, 127–28

Rosetta stone, 19

RP. *See* readiness potential

Russell, Bertrand, 29, 183

Ryle, Gilbert, 76

Sagan, Carl, 117

Sapolsky, Robert M., xvii, 66

Sartre, Jean-Paul, 34, 124

satisfaction resonance, 233–34

Schiller, Friedrich, 215

Schopenhauer, 116, 118

Schrödinger equation, 146–47
science, 228; brain understanding and, 109; consciousness and, 168; criterion for truth, 91–92; experimental research on consciousness, 121; funding and, 155; humanities versus, 183–85; Kant motto followed by, 92; new physics in light of history of, 154–55; philosophy versus, 85, 182–83; predictability and, 21, 135; public data versus data within, 251n7; as secondary to philosophy, 183; soul described by, 78–79
Science, 251n7
Searle, John, 125
Sebald, W. G., 44–45
self-awareness, 118, 131, 180; attempts to quantify, 223–26; consciousness measurement and, 221; in robots, 126
self-quantification movement, 40
Semmelweis, 120
sensory cortex, uninformative mode of, 140–41
sentient beings, 223
Seth, Anil, 13, 14, 131, 200
Shaw, George Bernard, 86, 106
simultaneous localization and mapping (SLAM), 228
The Singularity Is Near (Kurzweil), xv
SIT. *See* systematic inventive thinking
skepticism, 89–91; absence of, 188–89; happiness as requiring, 197
SLAM. *See* simultaneous localization and mapping
smart home, 39
smoking, Bernays use of predictability for ads of, 37–38
Snow, C. P., 183

social media, 39, 230–31; predictability and, 11
social pyramid, of predictability, 190, 208–9
society, less predictable, 206–7; hidden predictability in, 208–9
solar wind, 101
soldiers, 186, 188, 190
Solomon, Sheldon, 94
SOR. *See* stimulus-organism-response
soul, 135; evidence for, 78–81; futurists on, 79–80; idea evolution, 77–78; metaphysical notion of, 76; religious and philosophical references to, 76–77
The Soul Fallacy (Musolino), 79
Soviet Union, 186
speed of light, 104
Spencer, Percy, 30
spiritual enlightenment, 205
Spotify, 75
Stanford Encyclopedia of Philosophy, 65–66
Star Trek, 79–80; Data (robot) in, 125–26
statistical mechanics, 64
statistical theory. *See* randomness
sterilization, 215
Stevens, Michael, 62–63
stimulus-organism-response (SOR) theory, 71
string theory, 160
subconscious: death denial and, 95; determinism and, 69–70; as random number generator, 164
subjectivity, comfort zone and, 87–88
supercomputers: brain analysis by imaginary, 46–48, 53, 55; brain

noise not predictable by, 136; brain simulation, 148–49
superposition principle, 146–47; decoherence and, 157–58
superradiance effect, 144
survival, physical: parents survival and, ix–x; personal experiences of danger and, x–xi
Sutherland, Stuart, 221
synapse: cleft, 217; computational power in, 159
Synge, John, 106
systematic inventive thinking (SIT), 201

technology, evolution pace of, 116
Tegmark, Max, 143, 243n15
terror management, 94, 96
TestColor, 230
Thomson, G. P., 101
Thomson, J. J., 101
Thorndike, Edward, 71, 164–65, 194
thought experiment, 20; brain in glass vessel, 109–11; of Einstein, 103, 107; first step in, 109–11; QFT as born from, 156; question for, 115–19; of Wittgenstein, 122, 224
thoughts, particles giving rise to, 46
three-body system, 53, 161
time, red shift and, 107. *See also* atomic clocks experiment
TMS. *See* transcranial magnetic stimulation
Tononi, Giulio, 225
totalitarian principle, 90
Total Recall (Bell), 40
transcendence, 21
transcranial magnetic stimulation (TMS), 225

Trump, Donald, 11
truth: as hard for humans, 89; next door to, 28–29; objectivity and, 88–92; reservations about, 28–30; scientific criterion for, 91–92
TsungDao Lee, 91
Turin, Luca, 54
Turing Test, 125
The Two Cultures (Snow), 183
two-layer model, for brain outcomes, 163–68, 174

Ubermensch (beyond-man), 215
UBNR. *See* unpredictable but not random
ultraviolet catastrophe, 145
uncertainty principle, 50–51, 52; brain and, 148
unconsciousness, 122, 222
universities, predictability theses in, 41
unpredictability: FAPP and, 136, 142, 170–72; laws of nature and, 175; quantum randomness and, 142–43
unpredictable but not random (UBNR), 130, 131, 153; middle ground of, 212; new life definition as, 180; two-layer model and, 167–68
user preferences, 10

value alignment, 231
Vietnam War, 189
VisualDNA, 230
Voltaire, 216
Voyager, 9, 52, 116–17

Wallace, Alfred Russel, 93
Wall Street, 190
Was Einstein Right? (Clifford), 106–7

Weber, Max, 29, 182
Westworld, 125
White, T. H., 90
Wigner, Eugene, 158
Wikipedia, 156, 229
will, brain and, 219. *See also* free will
The Will and Its Brain (Kornhuber/Deecke), 74
Wilson, Robert, 30–31
Wittgenstein, Ludwig, 29, 98, 183; thought experiment of, 122, 224

women: Bernays smoking campaign and, 37–38; IG Farben experiment on, 118

Zeilinger, Anton, 66, 149
Zeno effect, 155
Zen story, 98
Zhang, Wanyi, 228
zombies, x, xi–xii, 18, 184; measurability of zombiness, 32–34; unreal Nikon camera experiment and, 31–32